MW00328029

Kidney and Pancreas Transplantation

Detailed Surgical Procedures and Management Protocols
2nd Edition

Ingemar J.A. Davidson, MD, PhD, FACS

Renal/Pancreas Transplant Department
Medical City Dallas Hospital
Dallas, Texas, USA

LANDES
BIOSCIENCE

AUSTIN, TEXAS
U.S.A.

VADEMECUM
Kidney and Pancreas Transplantation-2nd Edition
LANDES BIOSCIENCE
Austin

Copyright © 1998 Landes Bioscience
Printed in the U.S.A.

Please address all inquiries to the Publishers:
Landes Bioscience, 810 S. Church Street, Georgetown, Texas, U.S.A. 78626
Phone: 512/ 863 7762; FAX: 512/ 863 0081

ISBN: 1-57059-646-8

Library of Congress Cataloging-in-Publication Data
CIP applied for, but not available at time of publication.

Dedication

This book is dedicated to Robert McClelland, MD, FACS and his tireless devotion to surgical residency training. Dr. Mac is truly an extraordinary surgical personality. Having had the opportunity to visit numerous countries and institutions and having encountered a myriad of physicians and surgeons, I can assure the reader that there are few, if any, surgeons that can match the many qualities that he possesses. First off, there are few physicians that could equal his vast medical knowledge; in addition, he possesses a truly superb way of teaching and sharing his knowledge that is unique in the medical field.

Most surgeons know of him from the monthly publication, *Selected Readings in General Surgery* that he initiated, developed and, to this day, still edits. To those of us who have had the good fortune to work with him, other unique and "human" qualities are also apparent. I think that the best way to describe Dr. Mac is with words such as gentle, amicable, honest and fair, and his toughest judgments are passed on himself. These are qualities that many of us struggle to achieve. Qualities that, in Dr. Mac, are inherent and as much a part of him as are his Southern mannerisms and accent, which add a little spice to it all.

Dr. Mac is a person who truly fulfills the motto "service above all." Never have I witnessed, or for that matter, heard anyone say that Dr. Mac was irritated, abrupt or not willing to spend whatever time it took to explain a complicated problem. Typically, he will not give specific instructions, but rather, depending on circumstances, offer several options, leaving the final decision up to the student. These are the qualities of a real teacher.

The impressions he has made on literally hundreds of young training surgeons at Parkland Memorial Hospital with whom he has had personal contact is hard to assess or measure. The impact of *Selected Readings in General Surgery* on American as well as international surgeons is immense. Approximately two-thirds of the surgical residents in the US subscribe to this invaluable resource and Department Chairmen often use *Selected Readings* as a teaching tool in their surgical resident training programs. It goes without saying that Dr. Mac's influence on improved performance on the US Surgical Board exam is also one of his major contributions to the surgical profession.

Last, but not least, behind every good man stands an even better woman. Connie, I hope I didn't wake you every time I called your husband in the middle of the night asking for advice on a complicated Parkland trauma case.

Ingemar J.A. Davidson, MD, PhD, FACS
Editor

■ Contents

Editor

Ingemar J.A. Davidson, MD, PhD, FACS
Surgical Director
Renal and Pancreas Transplant Program
Medical City Dallas Hospital
7777 Forest Lane, Suite B-248
Dallas, Texas, USA
Phone: (972) 566-3959; Fax: (972) 544-3858
Chapters 1, 2, 3, 6

Contributors

Stephen T. Brown, MA
Medical Illustrator
Department of Biomedical
    Communications
University of Texas Southwestern
    Medical Center
Dallas, Texas
Chapters 2, 3, 4, 11

Carmencita Cava-Bartsch, RN
Transplant Coordinator
Renal and Pancreas Transplant Program
Medical City Dallas Hospital
Dallas, Texas
Chapter 6

Laura Coorpender, RN
Transplant Coordinator
University Transplant
    Program at Parkland
Parkland Memorial Hospital
Dallas, Texas
Chapter 5

Michael Edye, MD, BS, FRACS
Division of Laparoscopic Surgery
Mt. Sinai Medical Center
New York, New York
Chapter 4

Susan P. Graham, MD
Associate Professor of Internal Medicine
State University of New York at Buffalo
Medical Director
    of Heart Transplantation
Buffalo General Hospital
Buffalo, New York
Chapter 5

Ihsan Housini, MD
Assistant Professor
Department of Pathology
Parkland Memorial Hospital
Dallas, Texas
Chapter 8

Edwin H. Jenis, MD
Chairman, Department of Pathology
Millard Fillmore Health System
Buffalo, New York
Chapter 8

Mark Lerman, MD, FACP
Medical Director
Renal and Pancreas Transplant Program
Medical City Dallas Hospital
Dallas, Texas
Chapter 6

Lucille A. LoTempio, Pharm D
Pharmacy Department
Buffalo General Hospital
Buffalo, New York
Chapter 12

Christopher Lu, MD
Professor of Internal Medicine
Division of Nephrology
University of Texas Southwestern
    Medical Center
Dallas, Texas
Chapter 6

George L. Miller, MD
Associate Professor of Radiology
Director of Vascular Interventional
  Radiology
University of Texas Southwestern
  Medical Center
and Parkland Memorial Hospital
Dallas, Texas
Chapter 9

Carolyn E. Munschauer, BA
Graduate Student
University of Texas
  Southwestern Medical Center
Dallas, Texas
Chapters 6, 12

Karen Paolini, RN
Transplant Coordinator
Division of Renal/
  Pancreas Transplantation
Buffalo General Hospital
Buffalo, New York
Chapter 5

Arthur I. Sagalowsky, MD
Professor of Urology
Chief of Urologic Oncology
University of Texas Southwestern
  Medical Center
Dallas, Texas
Chapter 11

Anthony Setiawan, MD
Assistant Professor
Director, Ultrasound/Body CT
Parkland Memorial Hospital
Dallas, Texas
Chapter 10

Thomas Shanahan, PhD
Director, Histocompatibility Laboratory
Buffalo General Hospital
Buffalo, New York
Chapter 7

Maureen Ulrich, RN
Transplant Coordinator
Division of Renal/
  Pancreas Transplantation
Buffalo General Hospital
Buffalo, New York
Chapters 5, 6

Michael J. Wallace, MD
Assistant Professor of Radiology
University of Texas Southwestern
  Medical Center
Dallas, Texas
Chapter 9

Robert Waterhouse, MD
Assistant Professor
Department of Urology
Mt. Sinai Medical Center
New York, New York
Chapter 4

A Letter from the Editor

Dear Reader:

This publication represents the authors' current strategies based on the last several years of clinical and research experience, as well as literature review and attention to the experience of colleagues. The organ transplant procedure represents a unique clinical single-organ shock model. In the short term, the outcome depends on the quality of the donor organ and the effectiveness of perioperative management protocols, including surgical techniques and skill, fluid and drug management, organ preservation technique and ischemia times, to mention a few. In this respect, transplant procedures resemble those of shock resuscitation. Long-term outcome is more likely to reflect immunological factors, recurrent disease and patient comorbid factors. This book addresses peritransplant surgical management aspects in general, with emphasis on detailed intraoperative surgical technique. This is manifest in the detailed graphic representation that was materialized by many hours of communication from the surgeon's view to the illustrator's hand. When writing this book, the authors have mainly had the young training surgeon in mind; however, this handbook will prove a valuable resource for many, including operating room personnel, transplant coordinators and internal medicine transplant physicians. The authors realize that transplant science is rapidly developing and many management protocols may differ substantially between centers. To some extent, this book represents our slightly biased or personalized way of conducting transplant activities, or perhaps, what could be labeled as center-specific.

The early posttransplant period is commonly prone to problems requiring diagnostic and therapeutic procedures involving ultrasound, radiology and pathology departments, discussed in specific chapters. The new, exciting technique of laparascopic-assisted living related nephrectomy is also described in a separate chapter. The more extensive description of living related nephrectomy will be included in a separate publication on multiorgan procurement. One chapter is devoted to the description of peritransplant management and protocols. Again, the authors realize that transplant programs differ in style and protocols based on resources, hospital settings, support and research experience. Rather than literally accepting or copying our protocols, the reader should critically rewrite these to fit his or her program based on the many factors dictating a specific program's style. The message we wish to convey is that by using protocols and check lists the time saving is obvious and also safety and outcome markedly improved. As emphasized elsewhere in this book, the system approach to manage the transplant process is similar to that of the aviation industry.

Whenever possible, alternate commonly used methods or surgical techniques are mentioned or referenced. On that note, only selected literature is given after each chapter. This publication is not intended to replace or have a textbook type coverage, but rather as a practical hands-on guide for daily management issues. For future editions and publications, we would like to hear from you. Criticism, new techniques or alternate ways of handling difficult surgical problems are welcome. Should your suggestions be included in future publications, you and your institution will be appropriately recognized.

We look forward to hearing from you,

The Editor

◼ Foreword

Transplantation has evolved over the last decade at a remarkable pace. There are three distinct eras: precyclosporine, cyclosporine and the present era, with rapid development of new immunosuppressive agents. In each of these periods, a so-called "center-effect" has been observed when analyzing the overall results of each institution. Many attempts have been made to explain why a small number of centers have had results above the median. Although multiple factors impact outcomes, experience has taught us, as Dr. Davidson points out, that strict attention to detail and adherence to established protocols could explain this difference. Assuming, as I have, that this is a correct assumption, then incorporating the ideas, techniques and protocols found in this book, in addition to renewing our commitment to teamwork, will benefit us all, especially our patients.

The authors cover all aspects of transplantation in a concise but comprehensive manner. Attention is also paid to the establishment of pretransplant guidelines aggressively pursuing the transplantation of predialysis patients, which, surprisingly, is a concept not readily accepted among some of our peers. It is presented here without bias to specialty, but with genuine concern for patient outcome and cost effectiveness.

Even though immunosuppressive regimens have an inherent degree of variability, the fundamental concept remains the same. An area of emphasis in Dr. Davidson's book is the maintenance of adequate levels of cyclosporine and continuation of low dose prednisone. Over the years, the trend has been to minimize the dosage due to the economical load and experimentation with steroid withdrawal. This leads to a continuation of graft loss due to chronic rejection. The art of transplantation involves closely following the patient while optimizing immunosuppression. Dr. Davidson and his coauthors have also characterized the importance of teamwork between the transplant physicians and surgeons. This, among many other lessons, should be learned from this book.

The original intention of Dr. Davidson's book was to give the practicing physician and trainee an easy to follow, comprehensive guide to the science and art of transplantation. This has been accomplished without a doubt.

P. Vergne-Marini, MD, FACP, FACC
Medical Director, Transplantation
Methodist Medical Center
Clinical Professor of Medicine
University of Texas Southwestern Medical Center

Acknowledgments

The writing of this book has been greatly facilitated by several technical support personnel listed below. The editor would like to express a special appreciation for their efforts characterized by expressions such as "dedication, cooperation, going the extra mile, and professionalism." Thank you.

Christopher Candino, CST
Specialty Technician, General Surgery
Buffalo General Hospital Operating Room

Doug Hanes
Director of Medical Photography
Buffalo General Hospital

Barbara N. Heller
Secretary to Dr. Davidson
Buffalo General Hospital

Anna Towers, RN
Associate Coordinator
Buffalo General Hospital Operating Room

Kidney/Pancreas Transplantation: General Considerations and Current Issues

Ingemar J.A. Davidson

UNITED NETWORK FOR ORGAN SHARING

The United Network for Organ Sharing (UNOS) was established in 1977 to operate the National Organ Procurement and Transplantation Network (OPTN) and the National Scientific Registry for Organ Transplantation. In this role, UNOS functions as a contractor for the federal government and as a private, nonprofit corporation. UNOS is unique in that it is a private corporation which sets policy for a sector of the medical community and is subject to review and final approval by the Department of Health and Human Services. The goal of UNOS is to ensure equitable and efficient organ allocation for transplantation. The current objectives of UNOS are to:

- Develop policies for equitable access to available organs by those in need of transplantation and for the equitable distribution of procured organs.
- Establish and maintain standards of quality in organ procurement, distribution and transportation, histocompatibility testing and data collection and assure that such standards are met.
- Collect, verify, store, analyze and publish data concerning human organ procurement and transplantation.
- Provide information, communication and transportation systems which enhance the successful utilization of available donor organs.
- Increase the number of organs available for transplantation by providing information, consultation and guidance to persons and organizations concerned with human organ transplantation.

Kidney and Pancreas Transplantation, 2nd Edition, edited by Ingemar J.A. Davidson © 1999 Landes Bioscience

1

- Serve as a national resource regarding all aspects of organ procurement and transplantation.
- Provide administrative and logistical services which enhance the effectiveness of transplant professionals engaged in furthering transplantation.

UNOS includes every transplant center, organ procurement organization (organ bank) and transplant tissue typing laboratory in the United States. Members also include professional health organizations and voluntary members of the general public. UNOS is divided into 11 geographical regions with elected representatives or counselors from each. Various national committees of UNOS provide broad-based input into the 32 member UNOS board of directors which consists of 16 physicians and 16 lay persons. A national computerized waiting list for patients, linked to all organ procurement organizations and transplant centers, is maintained by UNOS. A point system has been established for the optimal distribution of organs. This system is based primarily on time on the waiting list, degree or quality of antigen matching and the panel reactivity antibody level, with pediatric recipients less than 11 years old assigned specific points. The point system may vary between UNOS regions and organ banks depending on local criteria.

Organs are first placed locally, then regionally and finally nationally. All blood type O kidneys are designated for blood type O recipients in order to avoid unfair distribution of O kidneys to non-O recipients. The final decision to accept a particular organ for transplantation will remain the prerogative of the transplant surgeon responsible for the care of the patient. There is mandatory national sharing for six antigen-matched kidneys, meaning that a patient identified through the UNOS computer as being ABO blood type compatible, with no HLA A, B or DR antigen mismatches for a specific organ, must be offered that donor kidney. The exception to this policy occurs when a kidney and a nonrenal organ (liver or pancreas) are to be transplanted simultaneously. Currently there is a "payback" rule for six antigen kidneys. UNOS maintains a scientific registry of data on all transplant recipients from the time of transplant through the loss of the graft or the death of the patient. The data includes immunosuppression drug use, first versus repeat transplant, recipient age, patient status and degree of histocompatibility at the time of transplant, organ preservation method and survival outcomes for graft and patient.

SHORTAGE OF ORGANS

Recent medical advances have increased the number of patients in need of transplantation to the point that the shortage of organs available for transplantation has reached alarming proportions. While the number of patients waiting for solid organs has increased in the United States in recent years, the number of organ donors has not. As of November 15, 1997, there were 55,994 patients on the UNOS waiting list for a solid organ transplant, the majority of whom (37,859) are waiting for a kidney transplant (Table 1.1). While the number of kidneys transplanted reflects organ availability, the increasing number of liver, pancreas, heart and lung transplants reflect improved medical technology, making these proce-

Table 1.1. *Solid organ waiting list as of November 15, 1997 and transplants performed in 1996**

	Wait List	Transplants 1996
Kidney	37,859	12,039
Liver	9,323	4,062
Pancreas alone	356	172
Kidney / Pancreas	1,596	850
Intestines	94	45
Heart	3,869	2,343
Heart / Lung	233	39
Lung	2,588	805
Total	55,994	20,354

* (UNOS Registry Data)

dures more feasible. This advancing technology is also reflected in the increasing number of patients on dialysis for end stage renal disease. In 1995, 68,870 new renal failure patients began dialysis in the United States, with more than 214,000 currently undergoing hemodialysis or peritoneal dialysis. In 1996, 1905 dialysis patients died on the waiting list.

THE END STAGE RENAL DISEASE (ESRD) TIMELINE

Many patients with increased creatinine are regularly seeing their primary physician or a nephrologist (Fig. 1.1, Phase 2). Interventions, such as aggressive blood pressure control, may delay or halt the development of progressive renal disease. As the creatinine rises to the 6-7 mg/dl range (GFR 10-15 ml/min) in nondiabetics, or 4-5 mg/dl in diabetic patients (GFR 15-25 ml/min) a vascular access should be placed since there is an impending need for dialysis (Fig. 1.1, Phase 3). Every effort should be made to allow 2 or more weeks before the use of a PTFE AV graft (i.e., GoreTex®, W.L. Gore, Flagstaff, AZ) and 4-6 weeks for a primary AV fistula to allow for maturation. However, about 75% of patients have no permanent vascular access by the time dialysis treatment becomes necessary (Fig. 1.1, Phase 4). Often, patients are referred too late and thus, require temporary central vein dual lumen catheters, with their inherent morbidity.

Depending on disease process and concomitant medical problems, many patients will remain on some form of dialysis treatment for the rest of their lives. Others are evaluated for a possible renal transplant and placed on kidney transplant waiting lists, where they may remain for varying periods of time depending on their blood type, preformed antibody levels, HLA matching, medical urgency and local organ bank efficiency (Fig. 1.1, Phase 4). The median waiting time to transplant for a blood type O patient is currently about 800 days (UNOS 1997). According to annual trends, of the more than 37,859 patients currently (11/97) on the waiting list only about 12,000 will receive a kidney transplant per year (Table 1.1). The transplant procedure includes organ procurement, in which the kidney is part of a multi-organ procurement effort (Fig. 1.1 Phase 5). Typically, the organs are excised within 48 hours after a donor has been identified. Ideally, the kidney will be implanted and reperfused within 20 h of excision (the cold ischemia time). The transplant surgical procedure itself takes only 3-4 h (Fig. 1.1, Phase 6), followed by 5-10 days in a hospital setting (Fig. 1.1, Phase 7). The posttransplant close surgical surveillance period lasts for 90 days, which is the timespan Medicare has set as the global fee coverage for surgeons (Fig. 1.1, Phase 8). During this time, immunosuppression should be designed to prevent acute rejection episodes (see chapter 6, Tables 6.10, 6.11, 6.13 and Figs. 6.2, 6.3 and 6.6). After 3-6 months the patients typically revert to their referring physician for follow-up, which lasts for the duration of the graft or the patient's life (Fig. 1.1, Phase 9). Graft failure, for whatever reason, brings the patient back to Phase 2, either as a hemodialysis patient or back on the waiting list for a repeat transplant. Patients may be placed on the waiting list when the graft is failing and perhaps be transplanted before the imminent need for dialysis.

A SINGLE-ORGAN SHOCK MODEL

The cadaver renal transplant (CRT) procedure is a unique, clinical single-organ shock model. The short-term outcome depends on the status and management of the organ donor, the duration of ischemia time and the circumstances surrounding the recipient's surgery. In this respect, the transplant operation and procedural measures resemble those of shock and shock resuscitation. Long-term outcome is thought to be determined by immunological factors and also greatly influenced by perioperative factors including intraoperative fluid and drug management, surgical technique and early postoperative management. We and others have previously reported poor outcome after CRT with the occurrence of delayed onset of urine output and delayed function (defined as the need for hemodialysis treatments) after transplantation. Therefore, identification of perioperative factors associated with this occurrence and institution of measures to induce urine flow immediately after surgery are exceedingly important to avoid hemodialysis and improve long-term graft function. This was also a major finding in the 1990 UNOS scientific renal transplant registry.

Another purpose of this chapter is to further highlight several peritransplant factors that affect outcome. These include, but are not limited to, perioperative fluid and drug management, donor and recipient selection and management, and ischemia time. An early immunosuppression regimen designed to ameliorate acute rejection is the major factor leading to long-term graft survival (see chapter 6, Figs. 6.2 a,b, and 6.6).

During the last several years a number of developments have affected the results of renal and pancreas transplantation. The decline in posttransplant morbidity and mortality has continued despite the expansion of eligibility criteria for trans-

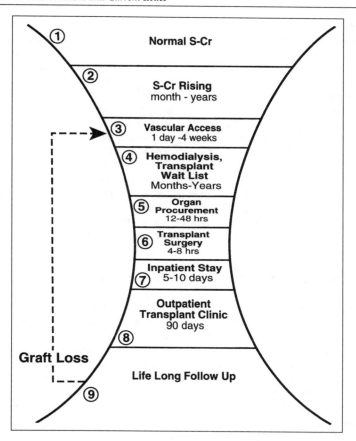

Fig. 1.1. The end stage renal disease (ESRD) time line as it pertains to renal transplantation. Patients with renal disease develop decreased renal function [2] and are followed early by their primary physicians and later usually by a nephrologist. As serum creatinine rises to 4-5 mg/dl in diabetics (GFR 15-20 ml/min) or 7-8 mg/dl in patients with no comorbidity (GFR 10-15 ml/min) permanent vascular access or peritoneal catheter is placed in anticipation of hemodialysis or continuous ambulatory peritoneal dialysis (CAPD) [3]. Patients may remain on hemodialysis for life or be evaluated for transplantation and placed on UNOS waiting list [4]. The transplant process involves donor maintenance and organ excision [5], recipient surgery [6], a short 5-10 day hospital stay [7], followed by close outpatient follow-up for 3-6 months [8], before reverting to their referring physician [9]. At any time patients may lose the organ to rejection or technical problems and revert to dialysis treatment.

1

plantation to include the higher risk recipients. Part of the increasingly successful outcome of transplantation is due to newer anti-rejection drugs. Specifically, the recent introduction of Mycophenolate mofetil (MMF) (Cellcept®, Hoffman-LaRoche, Inc., Nutley, NJ) has largely replaced azathioprine, and the new microemulsion formulation of cyclosporine A (CsA) (Neoral®, Sandoz Pharmaceuticals Co, East Hanover, NJ) with improved drug bio-availability resulting in less inter- and intrapatient variability, are likely to improve overall outcome. Perhaps of equal importance is the specific dose and combination of drugs over time, especially in the early days and weeks following transplantation. Consistently higher CsA blood levels of 300-400 ng/ml for up to 4 months (TDx/TDxFLx® cyclosporine Monoclonal Whole Blood Assay, Abbott Laboratories, Abbott Park, IL) has resulted in a predicted biopsy verified rejection rate of about 10% (see Fig. 6.7b, Table 6.15). The use of calcium antagonists, such as verapamil, induce higher CsA blood levels resulting in fewer acute rejections and partial protection from CsA induced nephrotoxicity. Our current drug protocols in the perioperative period are reported in detail in chapter 6 (Tables 6.3, 6.4, 6.5 and 6.10).

OUTCOME MEASURES OF KIDNEY TRANSPLANT

The outcome of transplantation is largely dependent on a number of system factors, as summarized in Table 1.2. Most outcome reports on renal organ transplantation are concerned with graft and patient survival, the most important factors from the patient's standpoint. The close relationship between early onset of urine output and improved outcome on each subsequent level cannot be overemphasized, and every effort should be made to institute early urine output. Immediate onset of urine production and large urine volumes are more beneficial than oliguria or no urine at all. Adequate renal function is physiologically more meaningful than just large urine volumes. Some renal function, even though not optimal but enough to avoid dialysis is preferable to delayed function necessitating hemodialysis in the early postsurgery phase. Delayed function offers a better prognosis for the patient than does a never functioning kidney, or graft loss, which is the worst scenario of the kidney transplant procedure short of loss of life. Clearly, from the patient's standpoint, these outcome variables are of increasing importance in the order that is listed in Table 1.2. From a study, research or statistical view the importance of variables is reversed. This is because the differences between study groups are more easily demonstrated for variables expressed by mean ± SD, such as urine volume and serum creatinine, than with variables described by percentages, such as graft and patient survival. This fact becomes obvious as survival variables exceed 90% and approach 100%, making it impractical or less meaningful to assess improved outcome measures.

A number of perioperative factors affect the outcome after renal transplantation (Table 1.3). The importance and application of these are reflected in protocol designs reported in chapters 5 and 6.

Table 1.2. Seven levels of outcome variables after kidney transplantation

1. Time of Urine Output Onset
2. Urine Volumes
3. Kidney Function
   serum creatinine
   glomerular filtration rate
4. Delayed Function
5. No Function
6. Graft Survival
7. Patient Survival

Table 1.3. System factors affecting transplant outcome

1. The effectiveness of national and local organ donor procurement program
2. Transplant program organizational structure and integrity
3. Perioperative fluid and drug protocols
4. Organ ischemia times
5. Immunosuppression drug protocols over time
6. Donor and recipient selection criteria

THE IMPORTANCE OF OUTCOME VARIABLE SELECTION

In the early days of transplantation, patient and graft survival outcome data were meaningful variables when survival percentages were in the 50-60% range. Currently, survival rates exceed 95% for patient and 90% for graft survival at one year. Therefore, these variables are less practical from a statistical standpoint to demonstrate improvement with a new treatment, i.e., an immunosuppressive drug. The incidence of rejection has also decreased in some centers to 20% or less within one year, making rejection a less practical method of measuring the outcome of transplantation. Therefore, renal function, including measurements of glomerular filtration rate and serum creatinine levels have become more important as outcome measures. Also, survival data has shifted emphasis from short- to long-term (5 and 10 year) survival. These changes in the way we assess transplant outcome are the result of tremendous progress made in the fields of immunology, drug development and surgical technique.

THE OPERATING ROOM

ANESTHESIA AND RECOVERY

General anesthesia with endotracheal intubation is the standard for the transplant procedure. Measures should be taken to maintain body temperature, including warming blankets. The authors prefer a supplemental epidural catheter

placed for postoperative pain control over approximately 24-48 hours. A nasogastric tube is placed after anesthesia induction and removed at extubation. A
percutaneous triple lumen venous central line is placed, preferably in the subclavian vein. An internal jugular line may be uncomfortable for the patient, since it
will be in place for the duration of anti-thymocyte globulin (Atgam®, Pharmacia/
Upjohn, Kalamazoo, MI) induction immunosuppression administration. The
authors also use one of the lines for intraoperative central venous pressure measurements to guide blood volume expansion (see chapter 6, Tables 6.4 and 6.5). In
programs where Atgam® is not part of the protocol, a central line may not be
necessary. It is beyond the scope of this book to discuss detailed anesthetic methods and drugs to be used. However, the authors cannot stress enough the importance of communication between the anesthesiologist and surgical team before
and during surgery as well as in the recovery room. There must be total cooperation and trust in the best interest of the patient, keeping in mind that short ischemia times mean better kidney function early after transplantation, as well as
improved long-term graft survival. In this context, it should be emphasized that
transplant patients, in addition to renal failure, often times have other significant
comorbidity including diabetes, hypertension and heart disease. Therefore, the
typical renal failure patient tolerates a lower dose of most drugs, especially seda-
tives. It is the authors' practice not to leave the patient until extubation has occurred and convincing, stable respiratory and hemodynamic situations are established after transplantation.

INSTRUMENTS AND EQUIPMENT
The authors cannot emphasize strongly enough the importance of being orga-
nized with regard to instruments and operating room equipment. When a transplant instrument cart was introduced the operating room time for simultaneous
kidney/pancreas transplant was shortened by approximately 2 h, from about 7 to
5 h. An example of a unicell storage device (Starsys, InterMetro, Wilkes-Barre, PA)
is pictured in Figure 1.2. This unit comes on wheels, is securely locked between
cases and is continuously restocked. The contents of the unicell may vary based
on basic instrument setup and the surgeon's personal preferences. The purpose is
to improve OR efficiency, time and cost, to decrease traffic in and out of the room
and to eliminate frustrating waiting time while specific items are located. In this
context, the availability of computerized surgical instrument tracking systems,
such as one designed by Dr. William Fry (Lynn, Ltd., 912 North Main #3, Ann
Arbor, MI 48103 (313) 996-1777) should be encouraged. In such a system, the
transplant unicell would be tracked by a simple code.

BACK-TABLE WORK
The back-table work occurs in two stages, as described in chapter 2. The first
stage is in the organ procurement operating room, immediately following organ
excision, and the second is in the recipient operating room immediately prior to
the transplant procedure. In both instances, optimal conditions should prevail.
The surgeons must sit down, have good lighting, use magnifying glasses and opti-

Fig. 1.2. A transportable instrument and equipment storage unit such as the one depicted in the figure results in shortened OR time, less traffic in and out of the operating room and elimination of much frustration.

mal instrumentation. In the procurement operating room the purpose is only to prepare the organs for packaging and shipment. In the case of kidneys, this involves separation of the en bloc removed kidneys, determination and documentation of anatomy, and perhaps performance of a biopsy. This procedure is described in more detail in chapter 2, and Figures 2.1-2.5. No reconstructive surgery should be performed in the procurement operating room. The procurement team is tired, and optimal surgical circumstances may not exist due to high traffic created from several teams and organ procurement personnel. These are factors that may result in sterility breaks. Furthermore, since the procurement surgeon may not be the transplant surgeon, he or she should not engage in any reconstructive surgeries which may not be in accord with practices of the transplant surgeon. Regardless of the circumstances, the procurement surgeon should not compromise on the quality of separating and assessing anatomy of the organs. Instruments for separating the kidneys should be easily available apart from those used for organ procurement. These must be kept sterile and moved from the main operating field to the back-table setting before contamination occurs.

The back-table work in the recipient OR may or may not require that the surgeon be seated. Often times the authors perform the final preparation with the kidney placed in a flat bowl of ice slush on top of the recipient's lower abdomen. This would be the case when the transplant surgeon also performed the procurement himself and does not expect any fine microvascular reconstructive surgery. However, when the kidney has been shipped in, or when reconstructive surgery is anticipated, a back-table setup should be in place. For pancreas transplantation, the back-table work is more elaborate, involving many steps and requiring microvascular instrumentation and special equipment (chapter 3, Tables 3.1-3.2, Figs. 3.1-3.4).

SPECIAL SURGICAL CONSIDERATIONS
The use of surgical magnifying lenses of at least 2.5X magnification is recommended. Technical errors are less likely to occur with proper magnification.

Micro-instruments
Several microsurgical instruments are highly recommended, available from companies such as Codman (Johnson & Johnson Professional, Raynham, MA) and Scanlan (Scanlan International, St. Paul, MN). For back-table work the authors prefer the nonlocking needle driver with a round handle that can be held as a pen and rolled in the surgeon's hand while suturing (Fig. 1.3a). The nonlocking needle driver allows continuous suturing without loss of eye contact with the operating field. Figure 1.3b shows a close-up view of the nonlocking needle driver used by the author for fine vascular surgery including backtable work and vascular access for hemodialysis. Needle holders with locking mechanisms often lead to loss of control while resetting the needle between each suture.

Along these lines the authors use micro-pickups of similar design depicted in Figure 1.4 a,b. As pointed out elsewhere in this publication, there must be no grabbing of blood vessels involving the intimal structures. Only pulling and pushing should suffice. Of course, perivascular adventitial fibrous tissue can be used to pull vessels.

The author recommends two types of micro-forceps. A small, indented "eye" improves the grip for vascular adventitial tissue. The sharp-tipped "Blue darter" forceps are especially useful for handling very small structures and for dilating vessels for corner suture placement (Fig. 1.4a,b).

For the transplant procedures themselves the same instrument principles prevail. However, the instruments must be longer to reach into the wound. Figure 1.5a

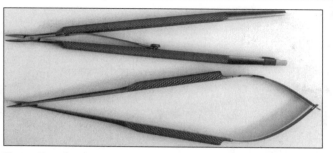

Fig. 1.3a. Examples of two types of micro-surgical instruments used for fine vascular reconstructions and back-table bench work.

Fig. 1.3b depicts a nonlocking needle driver in more detail.

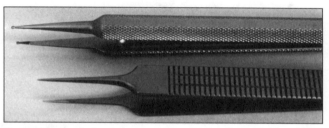

Fig. 1.4a. Example of micro-forceps, one with small indented eyes suitable to pick up small perivascular tissue and one with sharp pointed micro-forceps for very fine backtable work.

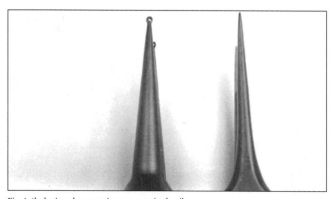

Fig. 1.4b depicts these two instruments in detail.

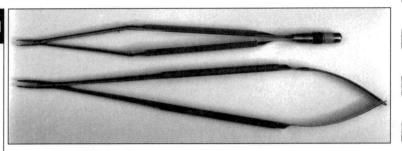

Fig. 1.5a. Examples of longer nonlocking needle drivers used for implanting the kidney and pancreas.

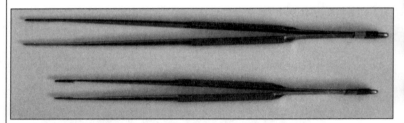

Fig. 1.5b shows long pickups for use in organ transplant procedures.

Fig. 1.6. Small vascular clamps (Heifetz) of different sizes, and shapes and angles that can be used for small vessel occlusions. Similar disposable plastic clamps of different occlusion pressures are also commercially available.

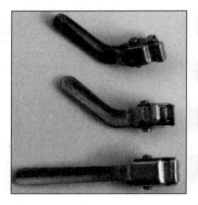

shows nonlocking needle drivers 26 cm long used for kidney and pancreas transplants. Figure 1.5b shows pickups suitable for organ transplantation.

Vascular Clamps

For backtable work the authors use metallic vascular clamps (Heifetz clips) of various sizes, configurations, and angles (Fig. 1.6), available from companies such as Weck (Edward Weck, Inc. Princeton, NJ). For larger vessels, the velcro bulldog clamps (Applied Vascular, Applied Medical Resources, Laguna Hills, CA) are preferred (Fig. 1.7a, b). These clamps are atraumatic and do not block the operating field (See Fig. 2.16, chapter 2). The velcro bulldog clamps come in two sizes and three different color-coded occlusive pressures. Currently the authors use the full pressure (white) clamps for both vein and artery. The more conventional larger custom-designed clamp on the external iliac vein may be more traumatic (Fig. 1.8).

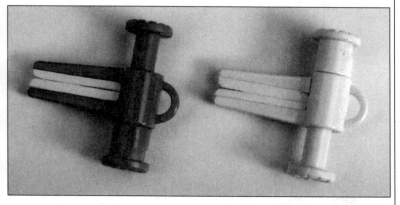

Fig. 1.7a. Color-coded, one-side velcro covered bulldog clamps, with full (white), half (dark blue) and 25% (light blue) (not shown) occlusion pressures.

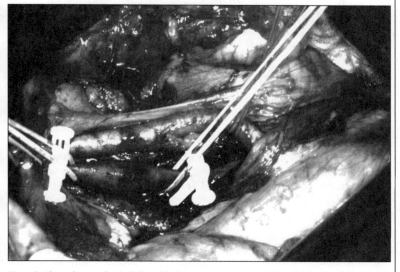

Fig. 1.7b. The authors prefer the full- and half-pressure types for occlusion of the external iliac arteries and veins for both kidney and pancreas transplants (also see Fig. 2.16, chapter 2).

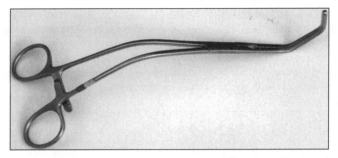

Fig. 1.8a. Example of a custom made standard vascular clamp developed for external iliac vein occlusion during kidney transplants.

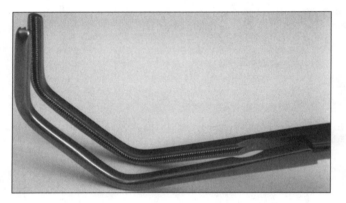

Fig. 1.8b. Depicts the details of this clamp.

Velcro covered clamps are also available in large conventional type. One of these is shown in Figure 1.9. An advantage to this clamp is that the shaft rests on the patients' abdomen and does not obstruct the operative field. Fine microscissors, such as Diethrich-Potts, are used to extend arteriotomoties in both renal and pancreas transplants (Fig. 1.10a,b). Round, tapered metallic perfusion cannulae, straight or angled, such as those available from Pilling (Fort Washington, PA), are quite useful for backtable organ flushing, as well as for local heparin flushing of vessels in the transplant situation (Fig. 1.11). Also, see Figure 2.13, insert b, chapter 2. The various micro-instruments, usually quite expensive, should be protected and stored in a padded box that is autoclavable (Fig. 1.12).

Fig. 1.9. (opposite, top) Example of a velcro covered atraumatic vascular clamp suitable for kidney or pancreas transplant. The angle and size is designed to fit a normal sized adult, in which case the shaft of the clamp will lie parallel with the abdomen causing no field obstruction, as is the case with clamps sticking out of the wound.

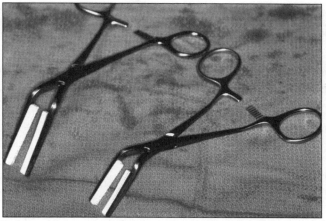

Fig. 1.9.

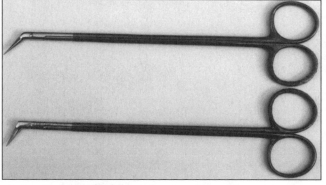

Fig. 1.10a. The Diethrich-Potts scissors shown here with two different angles are used to extend arteriotomies and venotomies as well as other precise vascular surgery.

Fig. 10.b shows these scissors in detail.

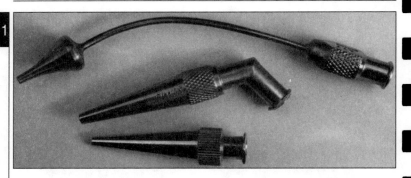

Fig. 1.11. Various types of perfusion cannulae used for organ perfusion and local flushing of iliac vessels during transplantation. (also see Fig. 2.13, chapter 2).

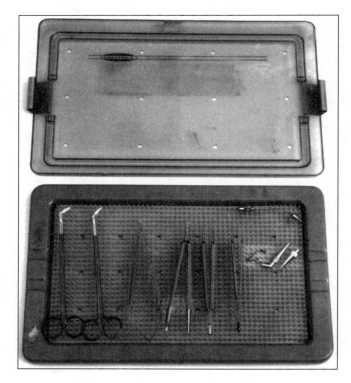

Fig. 1.12. This portable autoclavable storage deck protects and keeps vascular instruments assembled.

Sutures

Surgeons' choices of sutures and instruments are often based on experiences during training. It makes sense that a competent surgeon will know the sutures and instruments by name. The next paragraphs outline the authors' slight bias, sometimes held for good reason. For small vascular reconstruction, Prolene® sutures 6-0 or 7-0, 30" long, (C-1 or BV-1 needles) are used. Yet GoreTex® suture CV6 on a TT9 needle is often a better choice, the benefits of which are outlined in Table 1.4. The main advantage is the TT9 needle size, which is the same as the thread, eliminating suture hole line bleeding. This suture is used for the back-table pancreatic artery Y graft reconstruction (Fig. 3.4, chapter 3). The same GoreTex® suture on the larger TT13 needle is more suitable for recipient vascular anastomoses, both for kidney and pancreas transplants, and for closing kidney biopsy sites (chapter 2, Fig. 2.4b). The somewhat thicker thread on the GoreTex® suture tends to be less prone to cut through fragile tissues than the thinner Prolene®. The 6-0 Prolene® on a C-1 needle would be perfectly acceptable to replace the GoreTex® CV6-TT13 suture.

For closing the donor pancreatic duodenum over the staple line, the Prolene® 4-0 or 5-0 on an RB-1 needle is the author's choice. In cases of handsewn bowel anastomosis a two-layer closure with running 4-0 Vicryl® on a PS-2 needle is used in the mucosal layer, and 3-0 silk (control release) on an SH needle interrupted within the serosa.

Kidney transplant iliac fossa incisions are closed in two muscle-fascia layers with running PDS-II®, i.e., 0-0 or 1-0 on a CTX or CT-1 needle (See also page 50, chapter 2). Midline incisions for kidney alone or combined kidney pancreas transplants are closed with interrupted 0-0 Prolene® on a CTX (control release) needle. The authors strongly advise against running absorbable sutures such as PDS-II® for midline closures because of increased risk of wound dehiscence or later hernia formation.

Ureteral implants are performed with 5-0 PDS-II® on an RB-1 needle and the bladder closed in three layers with 4-0 or 3-0 PDS-II® on SH needles.

Subcutaneous sutures are probably never needed. Skin is approximated with staples. Table 1.5 summarizes suture choices used in transplantation.

Staple Instruments

It is the authors' opinion that stapling devices can be used in several transplant situations, especially during pancreas procurement and implant procedures. In the procurement situation a reloadable linear cutter stapler is used, such as the

---

Table 1.4. GoreTex® CV 6 TT-9 suture characteristics

- Needle the same size as thread
- Unique tying qualities
- Very strong
- White color shows well
- Requires 6-8 square knots

Table 1.5. Summary of sutures recommended in kidney and pancreas transplants

Bench Work	Suture	Size	Needle
Back-table	Prolene®	6-0	BV-1
	Prolene®	6-0	C-1
	Prolene®	7-0	BV-1
	GoreTex®	CV-6	TT-9
	GoreTex®	CV-6	TT-13
Recipient Vascular Anastomoses	GoreTex®	CV-6	TT-13
	GoreTex®	CV-6	TT-9
	Prolene®	6-0	CV-1
	Prolene®	6-0	C-1
Ureteral Implant	PDS-II®	5-0	RB-1
Bladder Closure	PDS-II®	4-0	SH
	PDS-II®	3-0	SH
Duodenum (Procurement)	Prolene®	5-0	RB-1
Handsewn Bowel Anastomosis			
Mucosa	Vicryl®	4-0	PS-2
Serosa	Silk	3-0	SH
Abdominal Closure			
Kidney (iliac fossa)	PDS-II®	0-0/1-0	CT-1
Midline	Prolene®	0-0	CT-1
	Nurolon®	0-0	CT-1 Control Release® "Pop-off"

TLC 55 or GIA 60 (Table 1.6) for dividing the duodenum both below the pylorus and at the level of Treitz ligament. Also, the authors staple the entire small bowel mesentery about 5 cm from the anterior surface of the pancreas using a linear stapler, TL 60 or TLH 90 (Table 1.6). The procured pancreas therefore has three staple lines (see pancreas back-table work chapter 3, Figs. 3.1 and 3.2). The multiorgan procurement procedure will be described in detail in a subsequent publication.

In the pancreas recipient, the distal (Treitz ligament) portion of the duodenum will be further shortened using the TLC 55 stapler. In cases of enteric drainage, which is the author's current choice, a side donor duodenum to side recipient small bowel (ileum) anastomosis is performed with the TLC 55 stapler (chapter 3, Fig. 3.11). In the case of urinary bladder drainage the authors have used the circular stapler (i.e., the CDH 25). Different sizes may be used depending on the lumen size of the duodenum (see chapter 3). Table 1.6 summarizes the most commonly used staplers by manufacturer.

Table 1.6. Stapler conversion chart for commonly used stapler devices in pancreas transplantation

U.S. Surgical [1]	Proximate® Ethicon (Endosurgical) [2]
GIA GIA 60 GIA 80 Reload GIA 60 Reload GIA 80	TLC TLC 55 TLC 75 TCR 55 TCR 75
TA	TL (RL)
TA 30 TA 55 TA 90 Reload TA 30 V Reload TA 30 Reload TA 55 Reload TA 90	TLV 30 TL 60 TLH 90 TRV 30 TRH 30 TRH 60 TRH 90
EEA	ILS (No reload)
CEEA 21 CEEA 25 CEEA 29 CEEA 33	CDH 21 CDH 25 CDH 29 CDH 33

1. U.S. Surgical, Norwalk, CT
2. Ethicon Endosurgical, Johnson & Johnson Inc, Somerville, NJ

SELECTED MANAGEMENT ISSUES IN KIDNEY/PANCREAS TRANSPLANTATION

BLADDER VS. INTESTINAL EXOCRINE PANCREAS DRAINAGE

In the United States, 92% of all pancreas transplants have been drained to the bladder. Early poor experience with intestinal leaks, sepsis and mortality have led to surgeon's reluctance to perform intestinal drainage. In addition, bladder drainage allows for monitoring of urine amylase to guide the clinical diagnosis and therapy of pancreatic rejection. The bladder drainage procedure is also technically attractive, avoiding possible delay of return of intestinal function posttransplant. However, enteric drainage has recently become more popular, with 15% of all pancreata drained to the intestine from 1994-96 (Newsletter, International Pancreas Transplant Registry April 1997, Fax (612) 625-2165; IPTR @lc.umn.edu). The following details the authors' experience in 39 simultaneous kidney/pancreas transplants, the first 22 of which were bladder drained and the last 17 enteric drained (Table 1.7). The lack of urologic and metabolic complications in the enteric drained pancreas is striking, especially in view of the lack of infections, intestinal leakage and readmission for metabolic acidosis and dehydration, which represent a marked cost saving. The quality of life improvement with enteric drainage is dramatic as compared to bladder drainage. It is the author's opinion that enteric drainage of the exocrine pancreas can be done without penalty of urologic and metabolic morbidity or surgical complications and graft loss. The U.S. 1994-1996 pancreas graft survival for bladder (n=1516) and enteric (n=221) drainage were similar, 82% and 77%, respectively. Most important, the kidney can be used as a marker for acute rejections in cases of simultaneous kidney/pancreas transplant. Factors contributing to improved outcome with the enteric drainage technique include higher CsA blood levels, which have reduced the incidence of acute rejection episodes to about 20% within the first year, improved pancreas organ procurement technique, the use of a simple side-to-side stapled duodeno-enteric anastomosis, and aggressive perioperative drug and fluid management of the recipient for improved posttransplant organ function. All these measures result in minimal need for posttransplant dialysis and have contributed

Table 1.7. Patient demographics and outcome variables for bladder (BD) and enteric drained (ED) simultaneous kidney / pancreas transplants

	BD	ED	P
	(n=22)	(n=17)	
Patient demographics			
Follow-up time(months)	48 ± 17*	22 ±11.6	p < 0.0001
Age	38 + 7	40.1 + 7	
Diabetes duration (years)	25 + 7	25.1 + 4.7	
Sex (M/F)	14/8	10/7	
CIT [kidney]	15.7 + 4.8	11.9 + 24.8	
HLA match [A,B,DR]	0.7 + 0.8	1.21 + 0.89	NS
Actuarial Survival: 1,2 and 3 year			
Patient	21/22 (95.4%)	16/17 (94.1%)	
Kidney	18/21 (86.3%)	17/17 (100%)	
Pancreas	16/21 (77.2%)	16/17 (94.1%)	
Complications			
Postabdominal abscess	1	0	
Wound infection	0	0	
Postop bleeding	0	1	
Pancreatitis	1	0	
Thrombosis	3	1	
Urine leaks	3	0	
Urinary tract infection	12 (55%)	0	
Urine retention	5	0	p < 0.001
Hematuria	5	0	
Bladder stone formation	2	0	
Acute rejections	5 (23%)	3 (17.6%)	
Readmissions (< 1 year)	21 (98%)	1 (5.8%)	p < 0.001

*SD

to improved outcome. Finally, octreotide acetate (Sandostatin®) administration, in the authors' opinion, is necessary to avoid intestinal distention posttransplant. However, in pancreas transplant alone, the kidney is not available as a marker for rejection. Therefore, bladder drainage may still be the best option in such cases.

Cyclosporine A (CsA) Blood Levels

CsA is part of the transplant immunosuppression regimen. There is little or no consensus as to what constitutes therapeutic or optimal CsA blood levels, which partly explains the variability of CsA assays and dosing schedules. Marked inter- and intrapatient variability are caused by several independent factors, including food and drug interaction and patient compliance. There is also controversy as to whether CsA therapy should be based on blood CsA levels or on CsA dose/kg basis. The recent introduction of a new microemulsion formulation of CsA (Neoral®), has simplified CsA therapy and monitoring. This new formulation, with improved intestinal absorption of CsA, is associated with more predictable and uniform blood levels, resulting in a consistent outcome. The authors have, since 1990, intentionally increased CsA blood (12 hour) trough levels for the first 16 weeks, targeting in the range of 300-400 ng/ml. From 16 weeks to one year, a level in the range of 200-300 ng/ml is the goal. These levels are approximately two times higher than previously customary in our cadaver kidney transplant program. Perhaps more importantly, the variation in CsA levels is less within the defined target window of 300-400 ng/ml. The goals of higher CsA blood levels and less variability are likely to be further improved with more widespread use of this new microemulsion CsA formulation (chapter 6, Fig. 6.3). In the authors' experience, CsA blood levels above 400 ng/ml prevent acute rejections after both simultaneous kidney/pancreas transplantation and cadaver kidney alone, but are associated with a high incidence (30%) of clinically significant renal toxicity. CsA levels between 300-400 ng/ml are associated with an incidence of rejection and toxicity of 6% and 10%, respectively. CsA levels below 300 ng/ml are associated with a greater than 20% incidence of rejection, but minimal (3%) incidence of toxicity. Our current higher CsA blood levels are possible because of the partial protective effects of the simultaneous administration of the calcium antagonist verapamil. The immunosuppression protocols are outlined in detail elsewhere (see chapter 6).

COMMON SURGICAL PROCEDURES IN ESRD PATIENTS

Vascular Access

Vascular access surgeries are the most common procedures performed in patients with ESRD. This subject was dealt with in a separate publication (see selected references).

NATIVE NEPHRECTOMY

Native nephrectomy was often performed prior to transplantation in the early days, but is seldom the case now. Unilateral nephrectomy on the transplant side at the same time as a kidney transplant is strongly discouraged. Elective native nephrectomy is indicated for medically uncontrolled renal hypertension, in cases of polycystic kidney disease to leave room for the transplant or when infection or bleeding into cysts are complicating factors.

TRANSPLANT NEPHRECTOMY

Transplant nephrectomy has become a fairly rare operation thanks to improved graft survival in recent years. The nature and indication, therefore, has changed markedly. While in the 1970s and '80s about 75% of these nephrectomies were done because of uncontrollable acute rejection before the CsA era, almost invariably in recent years transplant nephrectomies are done for chronic rejection. More than 50% of patients who lose kidney function and return to hemodialysis with discontinuation of their immunosuppression will ultimately require transplant nephrectomies. A mortality of 4% has been associated with retention of such grafts. The only significant problem encountered during transplant nephrectomy is the difficulty in controlling bleeding. It is the authors' opinion that transplant nephrectomies should be done by a surgeon experienced in transplantation and the transplant nephrectomy procedure. Therefore, patients should be referred back to their transplant center for transplant nephrectomy. With regard to technical detail for a transplant nephrectomy, the authors would like to emphasize the following points:

- The nephrectomy should generally be approached via the same wound through which the transplant procedure was performed.
- An extracapsular approach is used if the nephrectomy is performed within a few weeks after transplantation.
- An intracapsular approach should be used for kidneys that have been in place for a month or longer.
- The surgeon should dissect very close to the kidney, ligating and dividing the small vessels as they are encountered using nonabsorbable sutures.
- No attempt should be made to find and isolate the main renal artery or vein since this may lead to injuries of major iliac blood vessels.
- The ureter is isolated and oversewn as close to the bladder as practical, often leaving a short stump on the donor ureter itself.
- Perioperative antibiotic prophylaxis is used routinely.
- Drains should not be used. Every attempt should be made to ensure absolute hemostasis, as postoperative hematomas tend to cause infections and abscesses. Drains may, in fact, induce bleeding and infection.
- Patients who have been taken off immunosuppression and are back on dialysis sometimes present with acute rejection in a nonfunctioning kid-

ney with a tender, swollen graft. These patients should be "cooled" with 2-3 days of IV steroids, for instance methylprednisolone 250 mg IV.

INGUINAL HERNIAS
Inguinal hernias if small or asymptomatic should preferably be repaired at the time of transplant when the kidney is placed on the same side.

CHOLECYSTECTOMY
Cholecystectomy, preferably laparoscopic, is performed electively in symptomatic patients. Whether or not to remove all gallbladders with stones discovered incidentally on ultrasound is controversial. The authors' view is to only operate on patients who are symptomatic or have small, multiple stones or a dilated common duct.

HEMICOLECTOMY
Hemicolectomy for diverticulosis follows the same rule. Only patients with symptoms or a history of diverticulitis may need surgical intervention.

SELECTED READING
1   1997 Annual report of the US scientific registry for transplant recipients and the organ procurement and transplantation network—Transplant data 1988-1996. UNOS, Richmond, VA, and the Division of Transplantation, Bureau of Health Resources Development, Health Resources and Services Administration, US Department of Health & Human Services, Rockville, MD.
2   Ar'Rajab A, Dawidson I, Fabia, R. Reperfusion injury. New Horizons 1996; 4(2):224-234.
3   Burdick JF. What's New in Transplantation. J Am Coll Surg 1996; 182:170-76.
4   Cecka JM, Terasaki PI. The UNOS Scientific Renal Transplant Registry–1991. In: Terasaki PI, ed. Clinical Transplants. Los Angeles: UCLA Tissue Typing Laboratory, 1991:1-10.
5   Chiverton S G, Murie JA, Allen R D et al. Renal transplant nephrectomy. Surg, Gynecol, Obstet 1987; 164:324-7.
6   Darby CR, Cranston D, Raine AEG et al. Bilateral nephrectomy before transplantation: indications, surgical approach, morbidity and mortality. Br J Surg 1991; 78:305-7.
7   Davidson IJA, ed. On Call In Vascular Access-Surgical and Radiologic Procedures. Austin: RG Landes Co., 1996.
8   Davidson IJA, Munschauer CE. Albumin, the natural colloid: experimental data and clinical implications. In: Haljamae H, ed. Bailliere's Clinical Anesthesiology. London: Bailliere Tindall, 1997; 11(1):81-104.
9   Dawidson I, Ar'Rajab A, Dickerman R et al. Perioperative albumin and verapamil improve early outcome after cadaver renal transplantation. Trans Proc 1994; 26(6):3100-01.
10 Dawidson I, Rooth P, Lu C et al. Verapamil improves the outcome after cadaver renal transplantation. J Am Soc Nephrol 1991; 2(5):983-990.

11   Dawidson IJA, ed. Renal and Pancreatic Transplantation. In: McClelland RN, ed. Selected Readings in General Surgery. Dallas: The University of Texas Southwestern Medical Center at Dallas Publishing, 1993:20(7).

12   Dawidson, IJA, Sandor SF, Coorpender L et al. Intraoperative albumin administration affects the outcome of cadaver renal transplantation. Transplantation 1992; 53(4):774-782.

13   Phillips MG, ed. Organ procurement, preservation and distribution in transplantation. Second ed. Richmond: UNOS Publishing, 1996.

The Kidney Transplant Procedure

2

Ingemar J. A. Davidson, Arthur I. Sagalowsky
Illustrations: Stephen T. Brown

THE MULTI-ORGAN PROCUREMENT PROCEDURE

The kidney and pancreas organ excision from the brain dead cadaver donor is uniformly part of a larger multi-organ procurement effort. Due to the effect on seven human lives in one surgical setting, the multi-organ procurement procedure is perhaps the most important surgical intervention ever to be performed. Four people would actually die without the transplants (one heart, two lungs and one liver). Even though a kidney or pancreas transplant greatly improves the quality of life, these individuals will survive long-term with hemodialysis and insulin injections. Because of the magnitude of all aspects of multi-organ procurement, including both cadaver and open living related donor nephrectomy, this subject will be addressed in a subsequent publication. The new and exciting technique of laparoscopy-assisted living nephrectomy is described in chapter 4.

THE TWO STAGES OF BACK-TABLE WORK

The back-table work takes place at two stages and in two different settings. The first part takes place in the procurement operating room immediately after the removal of the kidneys en bloc and involves separation of the kidneys, documentation of anatomy and possible reperfusion with the UW solution and packaging. This part is described below and in Figures 2.1-2.4. The second back-table part occurs in the recipient operating room for each individual kidney, immediately prior to the transplant procedure (Figs. 2.5-2.9).

Kidney and Pancreas Transplantation, 2nd Edition, edited by Ingemar J.A. Davidson © 1999 Landes Bioscience

THE PROCUREMENT OPERATING ROOM

Immediately after the kidneys have been removed en bloc, they are placed on the backtable in ice slush (Ringer's Lactate or NaCl) solution. The overall anatomy is inspected. The authors prefer to place the kidneys in their normal anatomical position (Fig. 2.1). The ureters, attached to mosquito hemostats are also placed in their normal anatomical location. The en bloc kidney preparation contains the aorta, divided immediately above the superior mesenteric artery (SMA). In cases of simultaneous pancreas procurement the SMA has been divided next to the aorta. Occasionally, the origin of the SMA with a cuff of anterior wall of aorta is removed with the liver during the final excision in order to preserve a replaced

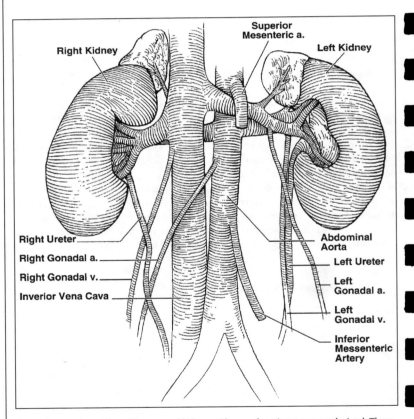

Fig. 2.1. An overview of the en bloc removed kidneys with normal pertinent anatomy depicted. The authors prefer to have the organs in their normal anatomical position for back-table work. The tip of the ureters should be attached to mosquito hemostats and positioned slightly downwards. The anatomy should be carefully assessed, including anomalies such as multiple arteries and veins and duplicated ureters. Size, measurements, precise anatomy, anomalies and possible injuries must be accurately documented by UNOS standards on forms supplied by the local organ bank.

right hepatic (accessory) artery. In such instances, great care is taken not to injure the renal arteries, which may originate at or only a few mm below the SMA. Distally, the aorta and the inferior vena cava (IVC) are divided at or just above, or at the level of their bifurcation below the distal umbilical tape tie around the aorta. The aortic perfusion cannula may have been brought with the kidneys to the backtable. The cannula is removed by cutting the two umbilical tapes with a knife. First, careful exploration is made for extrarenal arteries, especially lower pole arteries to the right kidney crossing on top of the vena cava. Rarely, a lower pole artery arises from the common iliac artery. Likewise, a renal artery may originate from the aorta above the SMA, or even the celiac artery, in which case injuries are common to the artery during the final organ excision. Multiple arteries occur in 15-20% of the population, with multiple veins considerably less common.

Left Renal Vein Dissection

The left renal vein is mobilized and divided next to the IVC (Fig. 2.2). No cuff of the IVC should be taken with the left vein, since this will make right renal vein reconstruction more difficult. Also, the left renal vein is long and needs no extra

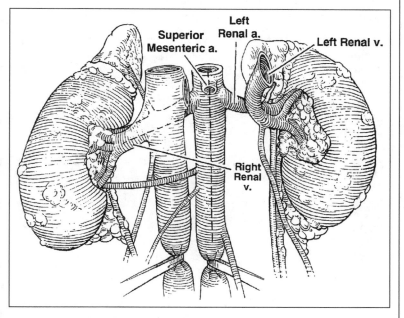

Fig. 2.2. The first step in separating the kidney is to divide the left renal vein from the inferior vena cava. The LRV is sharply dissected away from the underlying aorta, towards the left kidney. After the LRV is mobilized and placed towards the left, the anterior aspect of the aorta is exposed. The dotted line indicates the division line for the anterior aortic wall, preferably using the superior mesenteric artery (SMA) as a starting point shown in this picture. The aorta and the inferior vena cava (IVC) are divided above the umbilical tape placed during procurement. Great care is exercised not to injure accessory arteries between the aorta and the right lower pole kidney, as depicted in Figure 2.7.

2

length; in fact, it often has to be shortened. This will leave the right renal vein ready for reconstruction, depending on the transplant surgeon's preference (Fig. 2.9).

By mobilizing the left renal vein, the anterior aspect of the aorta is entirely exposed (Fig. 2.2). The authors prefer to divide the anterior wall of the aorta, starting at the SMA, using sharp scissors (i.e., Metzenbaum or Jamison) carefully dividing between the renal arteries, making sure that enough aorta is left on either side to create a patch for the arterial anastomosis. The entire length of the aorta is divided down to where the umbilical tape was tied.

Dividing the Aorta

Next, the back wall of the aorta is divided between the lumbar arteries (Fig. 2.3). At this point, the presence of multiple renal arteries is easily assessed.

As an alternative, some surgeons prefer to place the en bloc kidney preparation upside down and divide the back wall of the aorta between the lumbar arteries. This is perhaps slightly safer in terms of not inadvertently cutting close to the origin of the renal arteries, since the origin is slightly closer to the anterior aspect of the aorta. However, to the untrained eye, this approach may obscure the anatomy and potentially damage the left renal veins when the anterior wall is cut.

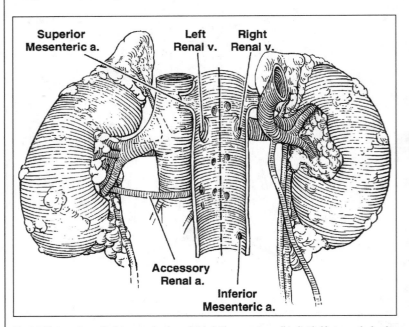

Fig. 2.3. The anterior wall of the aorta has been divided. The posterior wall is divided between the lumbar arteries, usually appearing in pairs. Accessory renal arteries (present in 15-25%) are carefully assessed (as shown here to the right lower pole). No further dissection should take place in the procurement operating room. The kidney anatomy is assessed and documented on forms supplied by the Organ Procurement Coordinator.

Organ Anatomy Documentation

The purpose of the back-table work in the procurement OR is 2-fold. First, the kidneys are separated. Second, individual renal anatomy is determined and documented accurately onto appropriate forms for the organ procurement organization (OPO) and the transplant center receiving the kidney. Any injuries or abnormalities must be reported in great detail. Should these occur, it is the author's opinion that they should NOT be repaired at this point, but rather precisely described with appropriate donor iliac arteries and veins sent with the kidney, for repair at the transplant site. If the procurement surgeon is also the transplant surgeon and one or two kidneys are used at the center, the procurement surgeon should accept the injured kidney for two reasons. First, this surgeon knows what injuries are present and should, therefore, be in the best position to judge the usability of the kidney. Second, if a procurement surgeon inflicted an injury, he or she should take responsibility by using the kidney at his or her own center. If a transplant surgeon does not think that a kidney is usable for anatomical/technical reasons, then the kidney should **not** be offered to others unless the exact anatomy and circumstances are reported.

Kidney Biopsy

The back-table work in the procurement operating room usually involves removal of some excess perirenal tissue. Sometimes, when the kidney quality is in question, as in the case of an older hypertensive donor, or in donors with marginal laboratory findings, a biopsy should be performed either using the needle or open biopsy (Fig. 2.4a). The authors prefer open biopsies in this setting, which gives a better sampling. The biopsy site should be the upper pole, away from the hilar region. The pathologist primarily will be examining the number and percentage of sclerosed glomeruli. Kidneys with 20-30% sclerosed glomeruli may not be usable. A frozen section biopsy can also be utilized to judge the degree of fibrosis. The biopsy, regardless of type, should also be processed for standard evaluation (chapter 8). The biopsy site is closed with a running permanent suture, such as GoreTex® CV6 on a TT13 needle (Fig. 2.4b).

BACK-TABLE WORK IN THE RECIPIENT OPERATING ROOM

Back-table Set Up

As with any fine, precise surgery, the back-table preparation requires the surgeon sit down with adequate light, microvascular instruments, complete sterility, operating room nurse support and, ideally, a competent assistant (Fig. 2.5a). Using strict sterile technique, the container with the kidney is opened. The table used should not have lower shelves at knee height, which make this delicate surgery painful for the surgeon's lower extremities. It is recommended at this point to take cultures from the perfusion solution surrounding the kidney. The authors prefer to place the kidney in a fairly flat rectangular bowl (Fig. 2.5b), containing slush ice on an OR lap, which stabilizes the organ. This set up is especially helpful should more extensive bench work be required, which is always the case with the pancreas portion (chapter 3 Table 3.1).

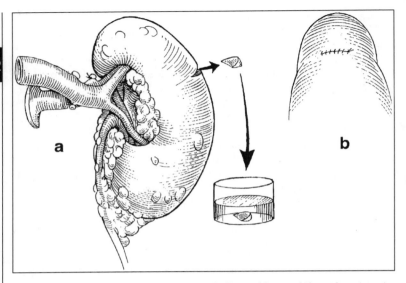

Fig. 2.4. When indicated, the authors prefer an open wedge biopsy of the upper kidney polar region, using a #15 knife blade. A 2-3 mm wedge, 5-7 mm deep, 10 mm long is excised (Fig. 2.4a). The defect is closed with a running suture, such as GoreTex® CV6-TT 13 (Fig. 2.4b). For handling of the biopsy see chapter 8, page 144.

If there are single vessels and no reconstructive surgery is anticipated, the authors prepare the kidney in a flat bowl with slush ice on top of the recipient's abdomen bypassing the back-table set up. Under no circumstances should this step be rushed to save time. Also, in the author's opinion, the kidney should always be inspected and prepared prior to skin incision. The left and right kidney usually require a different approach regarding the vein.

The Left Kidney
The Left Renal Vein
A mosquito hemostat is attached to perivascular tissue at the end of the vein to keep it slightly distended while dissection is carried out towards the hilum. Approximately mid-way, the adrenal vein is suture ligated towards the renal vein and divided a few millimeters away from the renal vein (Fig. 2.6). At this point, the adrenal vein can be followed and the adrenal gland itself excised by cutting next to the adrenal tissue, thereby avoiding inadvertent injury to a possible upper pole artery. At the lower border of the renal vein the gonadal vein will appear which may also be divided and suture ligated towards the renal vein. Some surgeons leave this vein intact for about 10 cm to preserve periureteral vascular integrity; however, the gonadal vein as it runs along or in proximity of the ureter, is easily excised. Care should be taken not to injure or inadvertently mistake the ureter for the vein. The renal vein should be followed to the level of the renal hilum, but not

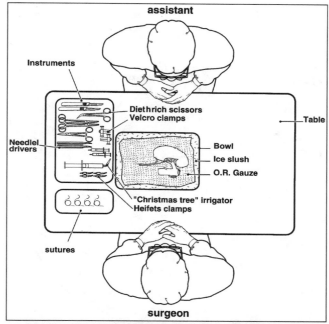

Fig. 2.5a. Back-table set up. The surgeon and the assistant must sit down with adequate light, magnifying glasses (x 2.5), micro-instruments and sutures such as GoreTex® CV6-TT 9, or Prolene® 6-0 or 7-0 on CV1 needles for any vascular reconstructive surgery. Examples of instruments are given in Figure 1.3-1.11, chapter 1. The kidney should be kept on ice slush in a flat bowl (Fig. 2.5b). For pancreas back-table instrumentation, see chapter 3, Table 3.1

into the hilum itself. It is preferable not to dissect too far, stopping 1-2 cm from the renal pelvis. Great care should be taken not to tear small venous branches from the renal vein. The authors prefer to occlude the end of the vein with a soft bulldog velcro clamp while expanding the renal vein with cold saline to make sure that there are no missed defects or branches, as shown for the right renal vein in Figure 2.9d.

2

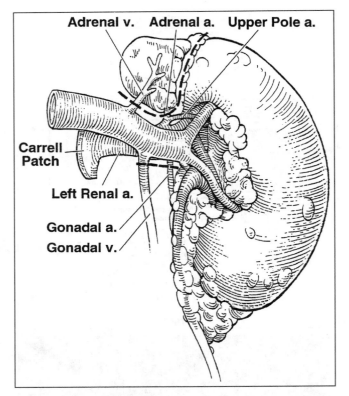

Adrenal v.   Adrenal a.   Upper Pole a.

Carrell Patch

Left Renal a.

Gonadal a.
Gonadal v.

Fig. 2.6. Back-table dissection of the left kidney involves: mobilization of the left renal vein, ligation of the adrenal and gonadal veins, removal of the adrenal gland, mobilizing the renal artery (LRA), ligating the adrenal artery and preparing the aortic cuff (Carrell patch). The adrenal artery often gives off an upper pole artery that should be preserved.

The Left Renal Artery

The left renal artery is usually attached to the long square aortic patch. If there is a single artery or two arteries close together, these are carefully dissected towards the hilum. Uniformly there is a branch to the adrenal gland, often arising from an upper pole artery. The adrenal artery is ligated, but the upper polar arteries must be preserved. Should a small upper polar artery inadvertently be injured or ligated, the kidney is still fine for use, but there will be a black area after reperfusion representing a small infarct, depending on artery size. The aortic patch can be created at this time or later, at the time of the implant. Suspicion of a missing lower pole artery (Fig. 2.7) should arise if a single main artery seems to be coming into the kidney a little too high, or is smaller than one would expect in relation to the kidney size. The authors have been involved in several situations

2

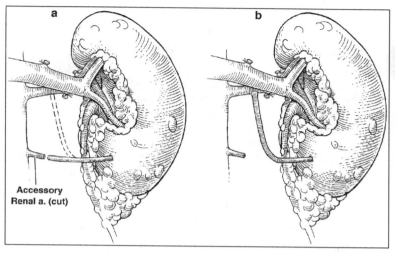

Fig. 2.7. Inadvertently cut left lower pole artery. The reconstruction method depends on local anatomy. An interposition graft to the larger upper main artery is the most practical method (dotted, inserted lines). Also see Figure 2.9e-f for additional reconstruction designs.

where a lower polar artery has been inadvertently cut. Depending on the anatomy and the length and distance between the arteries, reconstruction can be accomplished using different approaches (Figs. 2.7 and 2.8). In cases of dual renal arteries, different arterial reconstructive procedures are outlined (Fig. 2.8). In cases where one accessory artery (usually a smaller artery supplying the lower pole) is injured, an interposition vein graft should be used (Fig. 2.8f). The vein graft can be taken from any source, for instance the gonadal vein of the donor kidney itself. A missed and ligated lower polar artery may be disastrous, since this artery may also supply the ureter and parts of the renal pelvis. Reconstruction of this type will depend on remaining arterial vessel length size, and distances. This back-table reconstruction requires fine sutures (7-0 Prolene® (CV-1 needle) or GoreTex® (CV6 TT9), a calm environment, perfect lighting, magnifying glasses (2.5X) and competent assistance (Fig. 2.5a).

Right Kidney Preparation

The right renal vein usually requires reconstruction to gain length, as shown in Figure 2.9. The kidney should be firmly situated in a flat bowl in ice slush and immobilized by support from an OR lap to prevent it from floating freely (Fig. 2.5b). This makes suturing technically easier and more accurate. The IVC is totally freed and mobilized up to the right renal vein.

The Right Renal Vein Elongation

The proximal IVC and the left renal vein openings are connected and the combined defect trimmed as shown by the dotted line in Figure 2.9a, then closed by suturing the large opening which consists of the connected cava and the left renal vein (Fig. 2.9b,c). The authors prefer the GoreTex® CV 6 TT 13 suture for this

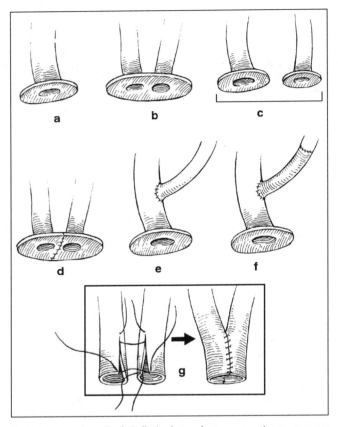

Fig. 2.8. Most common and technically simplest renal artery anatomy is one artery on an aortic Carrell patch (Fig. 2.8a), anastomosed to the external iliac artery. Equally simple are two arteries close together on one common aortic patch (Fig. 2.8b). Two arteries far apart can be anastomosed separately (Fig. 2.8c), or by combining into various types of reconstructions, as depicted in Figure 2.8d, e and f. Two arteries "fish-mouthed" together without an aortic patch are shown in Figure 2.8g. This scenario is identical to that of one artery without an aortic patch, which lends itself to an end renal artery to end hypogastric (internal iliac artery) anastomosis, which is preferred over end renal artery without aortic patch to side external iliac artery.

right vein elongation, due to the thick vena cava wall. By creating a pursestring effect, the angle between the right renal vein and the IVC is partially obliterated (Fig. 2.9d). Through this technique, the right renal vein can be elongated to the desired length. There are several lumbar branches, as well as the right gonadal vein that need to be suture ligated or closed. By applying a soft clamp on the renal vein close to the renal hilum, any missing branches can be detected by expanding the IVC and right renal vein with cold saline (Fig. 2.9d).

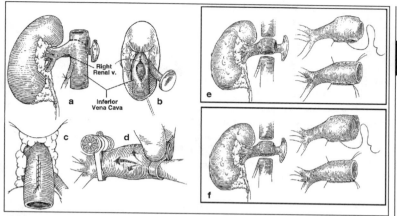

Fig. 2.9. The back-table work on the right kidney involves the right renal vein (RRV) elongation procedure, using the inferior vena cava (IVC). The opening between the IVC and the RRV is made along the dotted line (Fig. 2.9a), excising abundant venous tissue. Starting at the proximal corner, a running suture (i.e., GoreTex® CV6-TT 13) is used to create a water tight suture line (Fig. 2.9b and c). By creating a purse string effect, the angle between the RRV and IVC is partly eliminated. The suture line (and possible missed lumbar vein branches) is tested for leakage by expanding the vena cava and right renal vein with cold saline with a velcro bulldog clamp placed on the renal vein (Fig. 2.9d). Alternatively, the RRV is elongated by dividing the IVC below the renal vein (Fig. 2.9e) and suturing the IVC above and below the RRV. A variation of this procedure is shown in Figure 2.9f, where the front and back flaps of the cava are sutured from the RRV to the open corners of the left side of the IVC.

An alternative way to elongate the right renal vein is shown in Figure 2.9e. In this case the right renal vein ends at the very beginning of the left renal vein. By dividing and trimming the vena cava above and below the right renal vein, a continuous vein is created by closing/suturing the two flaps (Fig. 2.9f). The size of the venous anastomosis is guided by the size of the prepared cava flaps. These forms of caval elongation usually are superior from an esthetic perspective to that described in Figure 2.9a-d.

Anatomical Differences: Right vs. Left Kidney

The right renal artery and ureter are prepared similarly to that described for the left kidney. Yet there are several anatomical differences. First, the gonadal vein empties directly into the right side of the vena cava (Fig. 2.1) and is usually separated further from the ureter than on the left side. The adrenal vein also empties directly into the vena cava, therefore, there are no large named veins to ligate from the very short right renal vein. Even though the right kidney requires more benchwork and perhaps a somewhat more technically challenging venous anastomosis during the implant, in the author's experience, there is no difference in outcome in recipients receiving a right or left kidney. Because of these anatomical differences, the left kidney is preferred in the living related donor situation.

THE KIDNEY TRANSPLANT RECIPIENT OPERATION

SELECTION OF TRANSPLANT SITE

In an unaffected abdomen, the right iliac fossa is usually chosen for a first kidney transplant (Fig. 2.10, Incision A). For patients with a Tenckhoff catheter in place, the contralateral side is usually preferred. For second transplants, the left or contralateral iliac fossa is chosen (Fig. 2.10, Incision B). For third and subsequent transplants, an intra-abdominal approach is strongly recommended, which gives the option of choosing a side, depending on previously retained grafts and other anatomical vascular circumstances (Fig. 2.10, Incision C).

The intended skin incision is marked with a marking pen. The authors usually choose a curvilinear incision. While the initial skin incision is made with a knife, electrocautery is used through subcutaneous tissue, down to the external iliac muscle fascia. A starting point is obtained through the external iliac fascia and using a finger, the fascia is divided approximately 2 cm from insertion to the rectus muscle anterior fascia both upwards and downwards (Fig. 2.11a). Towards the pubic bone, the loose connective tissue attached to the lower portion of the rectus muscle is divided. Usually, the authors divide 1-2 cm of the tendinous portion of the rectus muscle insertion on the pubic bone to improve bladder exposure. The rectus muscle is attached again during wound closure. The area of the spermatic cord or round ligament is identified and the internal oblique and transversalis muscles are divided using electrocautery. With a finger under the transversalis fascia, the peritoneum is pushed medially to avoid inadvertent entrance of the abdominal cavity. Ideally, the peritoneum is freed from the posterior surface of

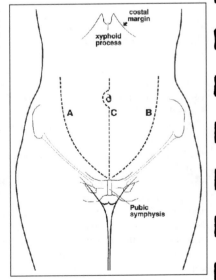

Fig. 2.10. For first transplants, the right iliac fossa approach is usually chosen (A) with the left side for second transplants (B). Third and subsequent transplants are better served through the midline intra-abdominal approach (C).

the transversalis fascia for about 2 cm to allow strong and adequate abdominal closure (Fig. 2.11b). The authors recommend a fairly long incision through the abdominal wall, which improves exposure. This mobilization of the peritoneum exposes the psoas muscle, the retroperitoneal space and the iliac vessels as well as the native ureter (Fig. 2.12). Approaching the lower end of the incision, the round ligament in females is ligated and divided with 2-0 silk, while the spermatic cord in males is surrounded with a vessel loop and preserved. Occasionally, the spermatic cord has to be divided to give adequate exposure. This may be the case after previous pelvic (hernia) surgery with adhesions and scar tissue. The inferior epigastric artery and vein are identified and divided en bloc or individually between clamps.

The authors prefer, at this point, to mobilize the bladder, previously filled with 150-250 ml of saline (depending on capacity) containing an antibiotic solution (cefalexin 0.5 g in 500 ml). The filling of the bladder facilitates its identification and mobilization. If the intravesical ureteral implantation is chosen, it is advisable at this point to start the midline cystostomy by incising the muscular layer exposing the bladder mucosa, but preferably not opening the bladder. This facilitates identification of the bladder at the time of implant. At this time, the clamp placed on the Foley catheter is removed and the bladder fluid drained, which increases exposure in the pelvis, facilitating the transplant procedure itself.

USE OF RETRACTORS

There are several very practical static retractors available. For many years, the authors have used the Bookwalter retractor, (Codman, Johnson & Johnson Professional, Raynham, MA). This retractor consists of a vertical arm connected to

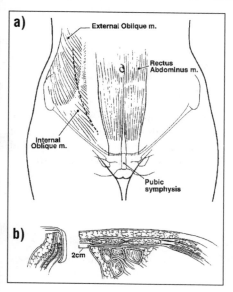

a)
External Oblique m.
Rectus Abdominus m.
Internal Oblique m.
Pubic symphysis

b)
2cm

Fig. 2.11. The curvilinear flank incision (Incision A in Fig. 2.10) is carried down through the external iliac fascia, approximately 2 cm from the rectus muscle fascia insertion line and then through the internal oblique and transversalis muscles and the transversalis fascia, pushing the peritoneum medially (Fig. 2.11a). A free posterior edge of approximately 2 cm should be prepared by freeing the peritoneum to facilitate abdominal wall closure and minimize the risk of stitching intra-abdominal structures (Fig. 2.11b).

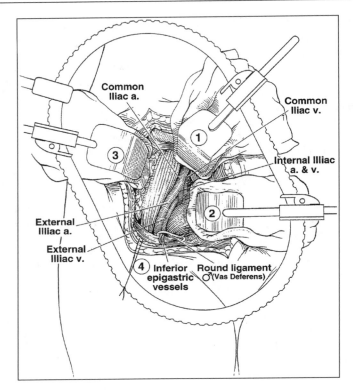

Fig. 2.12. The right iliac fossa surgical anatomy dissected and exposed during the transplant. The Bookwalter retractor is an excellent choice of static retractor device. The large
round or oval rings are preferred. Typically, one wide malleable retractor is used to retract the bladder and sigmoid colon (1) and a second blade (2) to retract the peritoneum
and small bowel towards the left upper quadrant. Additional blades (3) or sutures (4) are
used to enhance the exposure laterally and downward.

the operating table on one side and a horizontal arm on the other. This horizontal
arm connects to an oval or round ring providing the framework for retractors of
various kinds (Fig. 2.12). The typical setting and exposure of the transplant wound
is also shown in Figure 2.12. The authors use two wide malleable blades retracting
the peritoneum and the empty bladder medially. Also, one or two 0-0-silk sutures
are placed through the fascia and iliac muscles to pull the lateral and lower aspects
of the wound, thereby improving exposure.

Vascular Dissection
External Iliac Artery

The external iliac artery is mobilized first. The lymph vessels crossing the artery are ligated with 4-0 silk and divided. For the majority of cadaver renal trans-

plants with an available and usable aortic cuff, the external iliac artery is used as the anastomosis site. When there is no aortic cuff available, as is the case in living related transplants, the surgeon may choose to use the hypogastric artery for an end-to-end anastomosis to the renal artery. An end renal artery to side external iliac artery anastomosis without the cuff is technically more challenging and prone to vascular stenosis. These options are illustrated in Figure 2.13.

External Iliac Vein

The external iliac vein (EIV) is mobilized in a similar fashion. Distally, the circumflex iliac vein is a consistent vessel which runs across the artery in an upward and lateral direction. Ligation of this vein sometimes will improve the

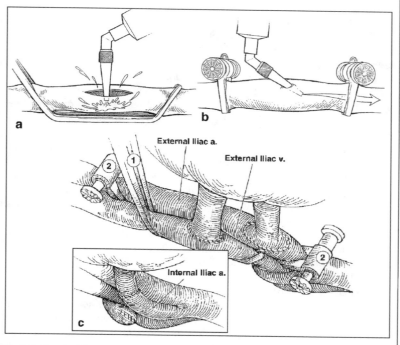

Fig. 2.13. Overview of anastomosis sites with single renal vessel anatomy. The end renal vein is anastomosed first to the side external iliac vein. The artery on the aortic patch is then sutured to the side of the external iliac artery. The kidney is wrapped in an OR lap covered with ice slush. The vessels are dropped through a small hole cut through the gauze. Custom made vascular clamps (1) or velcro bulldog clamps are used (2). An external iliac venotomy is made to match the renal vein (a). The EIV is irrigated. The end renal vein to side EIV is sutured with GoreTex® CV6 -TT 13 sutures, as detailed in Figure 2.14. The external iliac artery anastomosis is performed next. Before the arteriotomy is extended with microscissors, local heparinization is accomplished using an angled perfusion adapter attached to a 20 cc syringe (b). Heparinized saline is injected while the velcro clamps are sequentially opened. The authors do not use systemic heparin. In cases of no patch, or in a living related transplant, the arteriotomy is performed as an end-to-end hypogastric or internal iliac artery (see c).

exposure and length of the iliac vessels. During the dissection of the EIV, care should be taken not to tear small venous branches sometimes directed laterally into the psoas muscle or in the retroperitoneal space. If torn, such veins can cause severe bleeding which may be difficult to control. The authors prefer to ligate and divide these veins to increase the mobility of the EIV and avoid inadvertent tearing. Some surgeons divide and ligate the internal iliac vein when a short renal vein exists. Unlike pancreas transplantation, the internal iliac vein (hypogastric) generally does not need to be divided to improve exposure in kidney transplants. With proper procurement technique the right kidney should always come with the inferior vena cava, allowing a venoplasty elongation procedure as described in Figure 2.9.

VASCULAR ANASTOMOSES

The kidney is brought into the operating field resting in an OR lap with the vessels dropping through a small hole cut in the middle. The ureter and the lower kidney pole are directed towards the patient's feet. Ice slush is placed around the kidney to keep it cool at all times. This technique is identical to that used for the pancreas (chapter 3, Fig. 3.8). The authors prefer to perform the venous anastomosis first. A vascular clamp is placed on the renal vein. The authors sometime use a custom-made clamp (Fig. 2.13 and chapter 1, Fig. 1.8). Also, the velcro bulldog vascular clamps are excellent choices, both for the iliac vein and artery (Fig. 2.13 and chapter 1, Fig. 1.7). The velcro clamps are available in two sizes and with three increasing occlusion pressures (Applied Vascular, Applied Medical Resources, Laguna Hills, CA) (chapter 1, Fig. 1.7). The advantage of the velcro clamp is that no portion of it protrudes from the wound, allowing maximal exposure. The velcro clamp may also be less traumatic to the vessels than standard metallic occluding vascular clamps. A vascular instrument with velcro padded clamps with less damage to vessels is also commercially available (chapter 1, Fig. 1.9) (Applied Vascular, Applied Medical Resources, Laguna Hills, CA). These angled clamps lie along the abdomen, resulting in less obstruction of the surgeon's field than commonly designed clamps.

The Renal Vein Anastomosis

A small incision is made in the vein, large enough to accommodate an angiocatheter or an angled perfusion cannula attached to a syringe with heparinized saline. By injecting heparinized saline into the vein, it expands slightly and cleans any remaining blood from it. The venotomy is extended using Diethrich-Potts scissors to match the size of the renal vein (Fig. 2.13a). In case of a right renal vein elongated into the IVC, a larger venotomy is needed.

The corner stitches are placed in the iliac venotomy. The order and direction that the corner stitches are placed depends on local anatomy and the surgeon's preference. The authors prefer the GoreTex® CV6 sutures on the larger TT13 needles. The same CV6 size suture on the smaller TT9 needle, in the author's experience, is not practical for the vein or aortic patches; however, the TT9 needle is ideal for more delicate anastomoses, for instance, an end renal artery to end hypogastric artery anastomosis in a living related transplant. The advantages of these sutures are outlined in chapter 1, Table 1.4. Each inside corner suture is then placed inside out at the corners of the renal vein (Fig. 2.14a). The correct orientation of the renal vein is assessed using the previously tied adrenal and gonadal veins as markers. The corners are now tied while the assistant slowly drops the kidney into the wound to allow the renal vein to reach the anastomotic site (Fig. 2.14b). The corners are tied with 4 square knots, since both suture ends will be used. (The final knots will require 6 to 8 square knots.) A midway suture (Prolene® 6-0, CV-1) is placed on the opposite side, pulling the back wall away while the front side is sutured (Fig. 2.14b). Each side is sutured using a running technique. Ideally, one corner is run midway, to be met from the other corner and tied in the middle (Fig. 2.14c). This final midway knot is tied in 6-8 square knots. It is helpful to tilt the operating table as needed to increase exposure, both for the

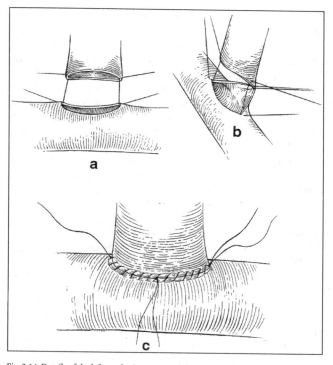

Fig. 2.14. Details of the left renal vein to external iliac vein anastomosis. Corner stitches (GoreTex® CV6-TT 13) are placed through the external iliac and renal vein (Fig. 2.14a). The corner stitches are tied in 3-4 square knots. A midway back wall retention suture (Prolene® 6-0, CV1) is placed which keeps the suture lines apart, facilitating exposure and exact suturing (Fig. 2.14b). Ideally, the first suture ends midway to meeting the suture sent from the opposite corner, tied with seven square knots (Fig. 2.14c). The back-wall suture is accomplished in the identical fashion, completing the anastomosis.

surgeon and the assistant. During suturing, the vein edges should be rolled slightly to allow intima to intima attachment. Grabbing the vessel with pickups is poor technique and lends itself to intimal injury, promoting thrombosis. Pickups are used to touch and push vessel walls, but not to grab. Exercising extreme atraumatic technique is advised. The authors prefer to use long, nonlocking needle drivers, which make suturing smoother and faster (chapter 1, Fig. 1.5a). Locking needle drivers tend to disrupt eye contact with the suturing field, since the needle needs to be reset between each stitch. The corner stitches and retention stitches are kept under constant tension by letting rubber shods hang outside the Bookwalter retractor, contributing to the ease of suturing. The venous clamp is kept in place while the arterial anastomosis is completed.

The Renal Artery Anastomosis

If not done previously, the Carrell patch is prepared by appropriate trimming of the aorta around the renal artery (Fig. 2.15a,b). In cases of multiple arteries, several options for back-table reconstruction of the renal artery anatomy are given in Figure 2.8. The single artery aortic patch is shown in Figure 2.8a. The two artery on the common patch reconstruction is illustrated in Figure 2.8b. The two artery on a long patch reconstruction is present with several alternative approaches (Fig. 2.8e,f). When there is no aortic patch available, as in the case of living related transplantation, an end renal artery to end hypogastric artery anastomosis is preferred, using a two or four corner stitching technique as described in chapter 3, Figure 3.3. Occasionally, a thromboendarterectomy of a stenotic plaque in the recipient at the bifurcation of the common iliac artery into the hypogastric artery is needed for a technically acceptable end-to-end anastomosis to take place. The end renal artery to side external iliac artery without a patch is technically more challenging and is associated with a higher incidence of posttransplant renal artery stenosis.

The suturing technique for the arterial anastomosis is similar to that described for the vein, however, the authors prefer not to tie one of the corners, usually the proximal, to facilitate exact suturing (Fig. 2.15). A midway retention suture is usually not needed. The lateral suture line is completed first, starting at the distal corner. Depending on local anatomy, vessel quality and the surgeon's preference, the suture may be run to the proximal corner. Alternatively, the proximal corner stitch may be tied and meet the distal suture halfway, as was the case in the venous anastomosis (Fig. 2.15c). The back wall of the arterial anastomosis is sewn from the opposite side with the two corner stitches meeting midway and tied with seven square knots (Fig. 2.15d). Before the last two stitches are placed, it is advisable to flush the anastomosis with heparinized saline to remove air bubbles.

REPERFUSION STEPS

Before the vascular clamps are removed, the authors prefer to make sure that the anastomosis is secure by applying a Heifetz clip or a velcro clamp to the renal artery and releasing the distal iliac artery velcro clamp (Fig. 2.16a). If there is significant bleeding or identifiable defect, this is addressed by a single stitch using GoreTex® CV6 TT-9 needle, or Prolene® 7.0 on a CV1 needle. Bleeding from needle

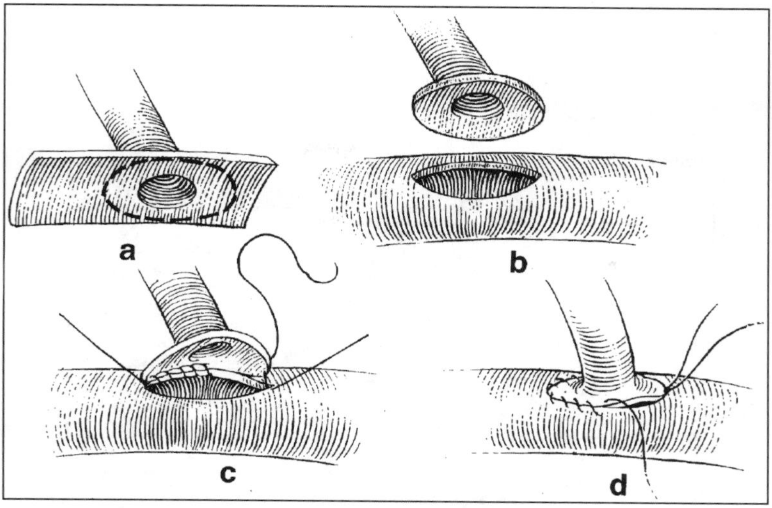

2

Fig. 2.15. The arterial anastomosis showing a single aortic patch-to-side external iliac artery (EIA). The creation of the oval patch, using fine Metzenbaum or Jamison scissors (Fig. 2.15a-b). The arteriotomy is made, matching the size of the aortic patch (b). Corner sutures (GoreTex® CV6-TT 13) are placed (Fig. 2.15c). The first distal corner is tied with 4 square knots, leaving the proximal suture untied but under slight tension (Fig. 2.15c). When the first running suture side is completed, the untied corner stitches are tied. One of these is tied to the just finished suture using 7-8 square knots (Fig. 2.15d). This technique, with the number of knots and suture arrangement is, in principle, identical to that of all anastomoses described in this publication. Before tying the second suture line, the anastomosis may be irrigated to remove air in the EIA.

holes in the external iliac artery will stop with a few minutes of compression. If there is no significant bleeding, the proximal external iliac artery clamp should now be removed.

Similarly, the vein may be checked with a velcro clamp applied close to the hilum, while the venous vascular clamp is removed (Fig. 2.16a). If two separate velcro clamps are used on the vein as shown for the pancreas transplant (see chapter 3, Fig. 3.8), the proximal clamp is removed first. Should the distal venous clamp be released while the proximal is still occluding, potential anastomotic rupture may follow due to high venous pressure build up in the femoral vein. Missed lumbar veins, usually on a right renal vein cava elongation in right kidneys, will require suture ligation. This situation should be the exception if the vein was properly tested on the back-table (Fig. 2.9d). The venous vascular clamps can, of course, be removed before the arterial clamps.

Normally, if exact suturing technique was used, extra stitches should not be needed. Again, oozing from needle holes is normal and stops within minutes by packing the wound with gauze or an OR lap. The authors warn against attempting to stop needle hole bleedings with extra stitches, as this will only increase bleeding

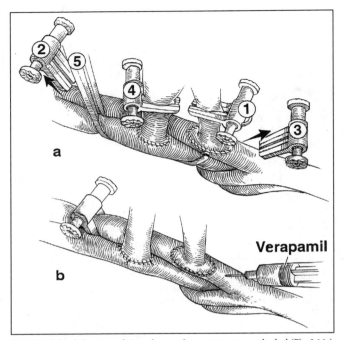

Fig. 2.16. Before kidney reperfusion, the vascular anastomoses are checked (Fig. 2.16a),
First, a Heifetz clip or a velcro bulldog clamp is applied to the renal artery (1), then the
distal velcro clamp (2) is removed followed by the proximal velcro clamp (3). A velcro
clamp is applied to the renal vein (4) and the specialty Satinsky clamp removed (5). Finally, the kidney is reperfused by removing the clamps on the renal artery and vein. When
all bleeding from the kidney has stopped, verapamil is injected into the external iliac
artery proximal to the anastomosis (Fig. 2.16b), while the external iliac artery is again
temporarily occluded with a velcro clamp. The verapamil injection is performed in increments of 2.5 mg up to 10 mg, watching the systemic blood pressure. The authors do not
use verapamil in cases of hypotension (less than 100 mmHg). Also, should systolic blood
pressure drop below 100 mmHg during verapamil injection, the injection is stopped (for
detailed intraoperative protocols, see chapter 6).

and potentially cause major technical problems with the anastomosis. Application
of a laparotomy sponge with slight pressure for a few minutes will uniformly stop
bleeding and prevent the surgeon from engaging in unnecessary stitching. In cases
where extra stitches are required, the authors prefer to use a Prolene® 7-0 or 6-0
on a BV-1 needle, or, a GoreTex® CV 6 on a TT 9 needle (Table 1.5). After hemostasis is obtained at the suture lines, the vascular clamps on the renal vessels are
removed to allow reperfusion. The kidney itself is now examined and significant
bleeders stopped, either by tying or gentle electrocautery, preferably bipolar. The
authors warn against using electrocautery close to or directly on the vessels or the
ureter. Neither should silk ties be used close to the ureter, but rather 5-0 absorbable material.

VERAPAMIL INTRA-ARTERIAL INJECTION

At this point, after hemostasis has been obtained, the authors inject verapamil into the renal artery, usually through the aortic patch, or into the iliac artery proximal to the anastomosis (Fig. 2.16b). In order to ensure delivery of the drug to the kidney, the external iliac artery is temporarily occluded distal to the anastomosis with the velcro bulldog clamp. Using a 30 gauge needle, 2.5 mg of verapamil is injected in increments up to 10 mg, while observing blood pressure changes. Usually, there is a 10-15 mm systolic blood pressure drop from this procedure. Should systolic blood pressure be 100 mmHg or below initially, the verapamil injection is not done. The benefits of the calcium antagonist is further discussed in chapter 6, Tables 6.7 and 6.8.

POSTPERFUSION RENAL BIOPSY

Some programs perform routine wedge or needle biopsies after reperfusion as part of clinical protocol. Others, including the authors, prefer biopsy, when clinically warranted, in the cold, bloodless kidney which makes suturing easy and exact with minimal bleeding after reperfusion (see Fig. 2.4). When performed after reperfusion, bothersome bleeding may occur requiring pressure in addition to suturing, adding significant operating room time. Baseline pathology is often useful later, too, for clinical management.

THE URETERAL NEOCYSTOSTOMY

THE INTRAVESICAL URETERAL IMPLANTATION

Before implanting the ureter and rearranging the retractors, it is wise to again ensure complete hemostasis around the kidney and hilum, as well as the anastomoses. Often, following vasospasm and cold exposure, blood vessels, when warm, tend to open and bleed. Once hemostasis has be re-established, the wide malleable retractors are removed and replaced with a bladder-blade retractor, retracting the rectus muscle laterally and downward and exposing the dome of the bladder (Fig. 2.17a). The previous median cystostomy is now identified and the bladder mucosa perforated with electrocautery or a tonsil hemostat. With the index finger inserted into the bladder, the cystostomy is further opened upwards and downwards using electrocautery. The cystostomy is enlarged just enough to place a gauze sponge at the dome end to insert a malleable retractor. This maneuver flattens the bladder opening and provides excellent exposure for the intravesical implant. Sometimes a retention suture at the kidney side of the bladder further improves exposure (Fig. 2.17a).

Next, the ureter is prepared, avoiding any rotation or twisting. The end of the ureter is spatulated for orientation (Fig. 2.17a insert), with the spatulation directed anteriorly in contrast to the extravesical procedure. At this time, the authors attach a mosquito hemostat to the ureter to avoid rotation. If the recipient is a male, the ureter must pass under the spermatic cord to avoid ureteral obstruction.

2

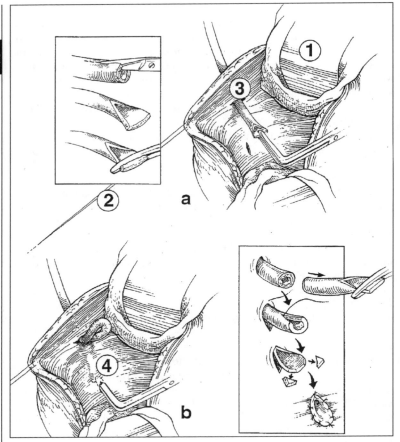

Fig. 2.17. The intravesical ureteral implant is performed through a midline cystostomy. Care is taken to align the ureter, avoiding rotation. The end of the ureter is spatulated (Fig. 2.17a insert), mosquito hemostat is attached to the end of the ureter to prevent rotation during the next few steps prior to implantation. A malleable blade attached to the Bookwalter ring is inserted in the bladder over a gauze in the bladder dome to retract and expose the inside of the bladder (1). A retention suture in the bladder may also improve exposure (2) (Fig. 2.17a). The classic Politano-Leadbetter implant uses two mucosal incisions in the bladder. The first is lateral (3), into which the ureter is pulled, then a 2-3 cm subcutaneous tunnel connecting this to the more medial portion (4), where the spatulated ureter is sewn (Fig. 2.17b insert). The authors, however, prefer to make the subcutaneous tunnel from the medial implant side laterally towards the bladder back wall using a right angle, and then perforate the bladder muscular layers with the ureter pulled directly into the implantation site. After the ureter has been pulled inside, it is further spatulated to the appropriate length and a corner stitch is attached to the mucosa (Fig. 2.17b insert). Two or three distal stitches also incorporate bladder muscle tissue to anchor the ureter and prevent it from pulling out at a later time. Three or four interrupted 5-0 PDS-II® sutures on each side complete the ureteral implant.

In females, the round ligament has been ligated and divided as previously described (pp 36). The traditional Politano-Leadbetter intravesical implant calls for two mucosal incisions with the creation of a submucosal tunnel between the intended orifice and the point from which the ureter is pulled into the bladder. This means that the ureter is pulled to its final position in a two-step fashion (Fig. 2.17b).

As an alternative, the authors prefer to pull the ureter through in one step. The native ureters are identified. Sometimes, the Foley catheter can be gently pulled up and out of the bladder to increase exposure. If feasible, the implant should be done approximately 1 cm from the trigone in the bladder floor. An opening is made in the mucosa approximately 1 cm long using low power electrocautery. Then, using a right angle hemostat, a submucosal 2-3 cm tunnel is created in the lateral superior direction. When the lateral mucosal fold has been passed, the clamp is passed through the bladder muscular wall using the surgeon's two left fingers to provide counter pressure outside the bladder. The assistant now places the prepared ureter, attached to a mosquito hemostat, into the right angle and the surgeon pulls the ureter into the bladder. Regardless of whether the ureter is drawn into the bladder and submucosal tunnel using one maneuver or two, the ureter must make a gentle angle through the bladder wall into the proximal end of the tunnel to avoid obstruction. The ureter is further spatulated, with care taken to leave appropriate length outside the bladder (Fig. 2.17b insert). This allows movement as the bladder empties and expands and accounts for the final position of the kidney in the wound. The ureter is sewn to the bladder mucosa using interrupted 5-0 PDS-II® suture on an RB-1 needle, starting with the corner stitch (Fig. 2.17b insert). The order in which the stitches are placed is a matter of local anatomy and the surgeons' choice. The authors usually place two distal corner stitches which include the bladder muscular wall as well as mucosa. These serve as anchoring stitches to prevent the ureter from being dislodged. Three or four interrupted full thickness stitches are then placed in the ureter on either side involving only the bladder mucosa. If there is no urine production, or there are questions about the smooth course of the ureter through the bladder wall, a #5 or #8 pediatric feeding tube is passed into the renal pelvis. If in doubt, 5-10 cc of saline without heparin, may be injected to verify free passage of fluid from the renal pelvis into the bladder. Usually there is no bleeding from the bladder. Minor oozing will stop after a few minutes of packing with sponges. At this point, flushing the Foley catheter tip with an angiocatheter will remove small blood clots, which may block urine passage immediately after surgery. A normal bladder is closed in three layers: first, a running 3-0 or 4-0 PDS-II® SH needle closes the mucosa to mucosa and ideally, the mucosa is inverted, which improves hemostasis and healing; second, a running 3-0 or 4-0 PDS-II® SH needle muscular layer and third, a serosal layer again with the same sized PDS-II® suture. With the use of this implant and bladder closing technique, the Foley catheter can be pulled as soon as 48-72 h after transplantation. The catheter is maintained longer in immobilized patients or while a urine output of ≥ 200 ml/hr persists. Sometimes, patients with small constricted bladders require longer catheter placement for both patient comfort and water-tight closure. As a general rule, no catheters or lines should be pulled without consulting the transplant surgeon. Another practical rule is not to remove catheters in late afternoon or evening hours when staffing is low and the team is outside the hospital.

THE EXTRAVESICAL URETERAL IMPLANTATION

The extravesical implant is preferred by many transplant surgeons. The ureter is sewn into a submucosal tunnel in the lateral part of the dome of the bladder from the outside. Even though the steps are similar to that of the intravesical approach, there are several important differences. First, the ureter is spatulated from the posterior view in contrast to the anterior spatulation utilized in the intravesical implant for approximately 10 mm. The final length of the ureter must be determined at this time. The bladder serosal and muscular layers are divided for approximately 5 cm in the lateral upward direction towards the kidney side (Fig. 2.18a) to create a mucosal trough as in the Lich-Gregoir reimplant. The mucosa is exposed and opened for about 5 mm (Fig. 2.18b). A proximal corner stitch is placed, as was the case for the intravesical approach, but now, the anastomosis is sewn from the outside (Fig. 2.18c). Next, two distal anchoring corner stitches are placed, both involving bladder muscle to prevent the ureter from dislodging. The two sides between the corners are then addressed. One side may be sewn from the inside (Fig. 2.18d). Depending on the surgeon's preference, the stitches, intermittent or running, are in close enough proximity to prevent urine leaks. Three or four stitches are also needed between the two distal corner stitches. The proximal corner stitch has to be passed between the bladder and the ureter, (if sewn from outside) to the contralateral side, to complete the other side of the anastomosis, which must also be sewn from the outside (Fig. 2.18e). Next, the muscular layer is folded on top of the ureter and continued for some distance beyond the anastomosis itself constituting the tunnel. It is hard to decide how tight the tunnel can be made. The seromuscular edges are closed over the ureter while a clamp presses the ureter downward toward the bladder lumen (Fig. 2.18f). In this way, the ureter remains free to peristalse within the tunnel. One or more sutures are placed between the ureter and the proximal end of the tunnel in the bladder wall to prevent ureteral retraction (Fig. 2.18g). At this point, the surgeon cannot test the patency of the ureter or inspect urine flow.

Some surgeons prefer slightly different approaches. One variation is a modification of the Barry reimplant. In this case, two transverse incisions are made. The ureter is passed between the two incisions, under the muscular layer, but outside of the bladder mucosa. This is achieved by passing an instrument (such as a tonsil hemostat) and pulling the ureter to the distal exit where the implant is performed as described above. This exposure at the implant site may be less optimal because of the transverse bladder incision compared to the Lich-Gregoir modification. Also, the tunnel is created blindly with some risk of inadvertently placing part of the tunneled ureter inside the bladder.

Advantages of the extravesical reimplants compared to the intravesical technique include less bladder dissection leading to less postoperative hematuria, clots and bladder spasm than the intravesical technique. Also, a shorter length of ureter

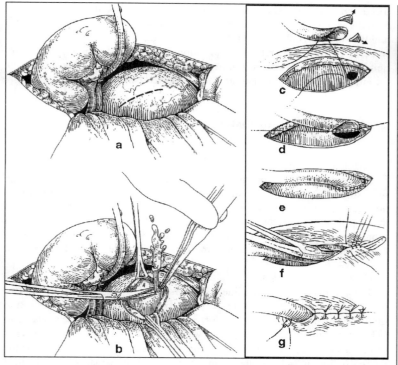

Fig. 2.18. The extravesical implant involves dividing the muscular bladder layer down to the mucosa for approximately 5 cm (the dotted line) (Fig. 2.18a). The mucosa is exposed to approximately 15 mm and opened (Fig. 2.18b). A corner stitch is placed from the spatulated ureter (Fig. 2.18c). Interrupted or running 5-0 PDS-II® stitches are then placed all the way to the distal end (Fig. 2.18d), where two or three sutures incorporate the muscular tissue to prevent the ureter pulling loose. The opposite side is sutured in a similar fashion, after the corner stitch has been pulled between the ureter and the bladder wall (Fig. 2.18e). The muscle and serosa layers are then sewn together over the ureter with a hemostat to ensure a loose enough tunnel (Fig. 2.18f). Finally, the ureter is attached to the bladder serosa to prevent it from pulling out (Fig. 2.18g).

is allowable with the extravesical technique. One disadvantage of the extravesical implant is that they are more tenuous and may require Foley catheter drainage for 6-8 days to minimize the risk of urine leak.

URETERAL STENTS

Some surgeons place ureteral drains, either through the urethra tied to the Foley catheter (outside the urethra) or through the bladder dome as a suprapubic tube. It is the author's opinion that ureteral stents are unnecessary in routine uncomplicated transplants. In fact, we believe these catheters may cause complications such as infection, urine leaks and hematuria, with kidney pelvis clot formation and obstruction. The overall rate of ureteral leak or obstruction in the author's experience at Parkland Memorial Hospital, Dallas, Texas between 1985 and 1990 has been only 2.4% (11/451) despite avoidance of routine ureteral stents. In the rare cases when ureteral complications do occur and reimplantation is required, ureteral stenting is recommended for maximal safety and healing of the secondary repair. (see also chapter 11 on urological complications.)

CLOSURE OF THE ABDOMEN

SURGICAL DRAINS
Before closing the abdomen, complete hemostasis is again ensured. Although some surgeons routinely place suction drains, the authors strongly advise against this practice. First, closed suction drains soon tend to be occluded by clots and therefore lose the intended purpose. Also, drains tend to induce bleeding from rubbing, or even worse, cause injury if placed against an anastomosis. Finally, the wound infection rate is higher when drains are left in place for 24-48 h or longer.

CLOSING TECHNIQUE
The authors close the curvilinear standard kidney transplant incision in two running layers using 0-0 PDS-II® on a CTX or CT-1 needle. A Prolene® suture would be equally efficient. Hernia formation, although rare, is likely related to technical error and/or inherent fascial weakness rather than the suture material used. The first and inner layer includes the transversalis fascia and muscle and the internal oblique muscle. Great care must be exercised not to inadvertently include peritoneum and the underlying bowel. The authors have been involved in two such cases in 15 years, resulting in one graft loss and one fatality. To avoid such a devastating error, the authors recommend separating the peritoneum off the back wall or the rectus muscle for about 2 cm, giving space for adequate suture bites during closing (Fig. 2.11b).

For adequate exposure and safe closure of this inner layer, the authors use two sutures starting in each corner and meeting at the midpoint, tied to each other. The second layer also utilizes a running PDS-II® suture for the external oblique muscle and fascia. Suturing the subcutaneous tissue or fat is not needed in most cases. The skin is closed with staples or any preferred suture technique.

SELECTED READING
1   Dawidson IJA, Rossi P, Meyer D. Multiple organ procurement protocol: details and variations of surgical technique. Dallas: University of Texas Southwestern Medical Center at Dallas Publishing, 1993:1-14.
2   Sagalowsky AI, Dawidson IJA. Surgical Complications of Renal Transplantation. In: Jacobson H, Striker GE, and Klahr, S, eds. The Principles and Practice of Nephrology. St. Louis: Mosby, 1995 (second edition); 822-26.

The Pancreas Transplant Procedure

3

Ingemar J. A. Davidson
Illustrations: Stephen T. Brown

BACK-TABLE WORK IN THE ORGAN RECOVERY OPERATING ROOM

If the liver and pancreas have been excised separately, there is no need for back-table work in the organ recovery operating room, beyond inspection and packing the pancreas. During the excision, some surgeons prefer to mark the splenic artery with a suture, such as Prolene® 6-0, making it easier for the transplant surgeon to identify. The spleen is left attached to the pancreas, to be removed in the recipient operating room. Small pieces of the spleen, however, may be taken and sent with the various organs for tissue typing and crossmatch purposes.

In cases of en bloc liver and pancreas excision, the organs have to be separated on the backtable before being packed and shipped. The magnitude of this back-table separation depends on the degree of dissection performed in the heart beating donor. It is the author's opinion that most of this dissection is best accomplished in the heart beating donor, assuming stable hemodynamics. In cases of a hemodynamically unstable donor, the need for cold perfusion and en bloc excision becomes a priority. The procurement surgeon must be familiar with variations in procurement technique. Detailed multi-organ procurement techniques and variations will be described in a subsequent publication.

BACK-TABLE WORK IN THE RECIPIENT OPERATING ROOM

The pancreas requires fairly extensive back-table preparation. The equipment needed for the back-table work is listed in Table 3.1. It is imperative that the surgeons sit down and are comfortable. It is very helpful to place the pancreas in a flat bowl with ice slush, with an OR lap to stabilize the pancreas and prevent it from

Table 3.1. *Suggested equipment for back-table work on the pancreas*

1 Flat bowl filled with slush ice
2 4-0 and 2-0 silk ligatures
3 Gore-Tex ® CV6 TT9 sutures (N=8) (chapter 1, Table 1.4 and 1.5)
4 Prolene® 5-0 sutures on RB-1 needles (N=2)
5 Mosquito hemostats with rubber shods (N=5)
6 Metzenbaum scissors
7 Diethrich-Potts fine vascular scissors (chapter 1, Fig. 1.10)
8 Nonlocking microvascular needle driver and pickups of type shown in Figures 1.3-1.4
9 Small vascular clamps (i.e., Heifetz clamps (Fig. 1.6) or velcro bulldogs (Fig. 1.7))

unnecessary movements. Both the surgeon and the assistant should use magnifying loupes of at least 2.5X magnification. One pair donor iliac arteries and veins must be available, and preferably packed with the pancreas, to avoid inadvertently losing this vital tissue.

The container with the pancreas is opened under sterile conditions. It is advisable to culture the perfusion fluid surrounding the pancreas as a guide for treatment should fever develop in the recipient's first few posttransplant days. However, in the authors' experience, positive cultures from the perfusion solution rarely can be related to postoperative infections in the recipient.

STEPS IN BACK-TABLE WORK

A summary of back-table pancreas reconstruction work needed is given in Table 3.2.

1) The spleen is removed by tying individual vessels with 2-0 silk ligatures only towards the pancreas. It is imperative that these are firmly tied since there is a tendency to bleed through these ligatures as the tissues warm up (Fig. 3.1, step 1). The tissue and vessels are cut on the spleen side of the tie. As an alternative, the spleen may be excised last, as it helps in stabilizing the organ and can be used as a handle. Some surgeons excise the spleen after reperfusion inside the recipient.

2) The distal portion (Treitz ligament) of the duodenum usually requires further mobilization and shortening by separating it from the pancreas. For this purpose, the authors use 4-0 silk ties (Fig. 3.1, step 2). The excess duodenum is removed using a stapler, such as TA-55, after the implant (Table 1.6). In cases of bladder anastomosis this excess portion of duodenum is utilized for insertion of the circular stapler prior to its removal (Fig. 3.10a).

3) The staple line across the small bowel mesentery is oversewn with 4-0 Prolene® (RB-1 or taper SH needle) running suture (Fig. 3.1, step 3). Even though the authors have had no technical problems with the stapling of the mesentery in 39 cases during the procurement with a TA-90 (Table 1.6, chapter 1), many pancreas transplant surgeons prefer individual tying of the mesenteric arteries and veins.

Table 3.2. Back-table pancreas bench surgery steps

1 Removal of the spleen (Fig. 3.1, step 1)
2 Further dissection between the pancreas and the duodenum of the LT portion, to be removed after reperfusion (Fig. 3.1, step 2)
3 Oversew the TA staple line of the small bowel mesentery (Fig. 3.1, step 3)
4 Running suture Prolene® 4-0 suture of the pylorus stapled and of the donor duodenum (Fig. 3.1, step 4)
5 Mobilization, as needed of the portal vein (Fig. 3.2, step 1)
6 Restoration of the arterial vascular anatomy using the donor iliac vessels into a Y-graft (Fig. 3.2)

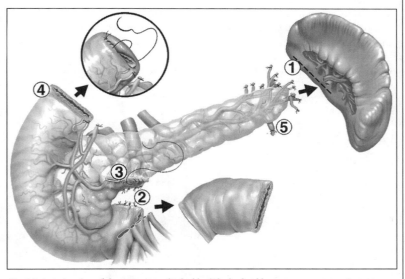

Fig. 3.1. Anterior view of the pancreas on backtable. (The back-table pancreas reconstruction steps are also described on page 52). First, the spleen is removed by individually tying (2-0 silk) arteries and vein along the dotted line (1). The distal portion of the duodenum is further mobilized and separated from the head of the pancreas (2). The removal of the excess duodenum is performed after revascularization, using a TLC or TCR 55 stapling device (chapter 1, Table 1.5). A running Prolene® 4-0 suture (RB-1 needle or taper SH needle) is sewn along the staple line across the stapled mesenteric suture line (3). The stapled duodenum, towards the pylorus side, is oversewn with a running 4-0 Prolene® suture (RB-1 needle) (4) (insert picture). If not adequately done during procurement, the inferior mesenteric vein may need to be retied or suture ligated (5). The order of these steps may vary depending on the surgeon's preference.

4) The pylorus side of the duodenum is oversewn with the staple line inverted using a running 5-0 Prolene® on an RB-1 needle. (Fig. 3.1, step 4). This may add extra security against urine leaks in cases of urinary bladder drainage.

5) The catheter used for portal vein cold perfusion from the inferior mesenteric vein (IMV) is removed and the IMV ligated at the inferior edge of the pancreas (Fig. 3.1, step 5). Of course, other entry ports may be used for portal vein

(PV) perfusion. In cases of pancreas transplantation the IMV site has the least risk for injury to the splenic vein (SV) or PV, assuming that the perfusion catheter is located only in the IMV.

6) Sometimes the PV requires a little further mobilization (Fig. 3.2). In the author's opinion, the PV does not require vein graft elongation. Optimal length is about 2-3 cm.

7) The arterial reconstruction is performed using a Y graft from the donor iliac arteries (Figs. 3.2 and 3.3). Usually, the donor hypogastric artery size will match to the splenic artery. The hypogastric artery also naturally makes an appropriate curve which further makes this artery fit well to the splenic artery. The technique used is a four corner stitch placement (Fig. 3.3) with Gore-Tex CV6-TT9 sutures. The four corner stitches are attached to rubber shods that loosely hang over the side of the flat bowl containing the pancreas in slush ice. This keeps the anastomosis open at all times during suturing which eliminates inadvertent back wall suturing. Also, the eight needles and suture threads will be lined up in an orderly fashion which will minimize confusion. One corner stitch is tied in

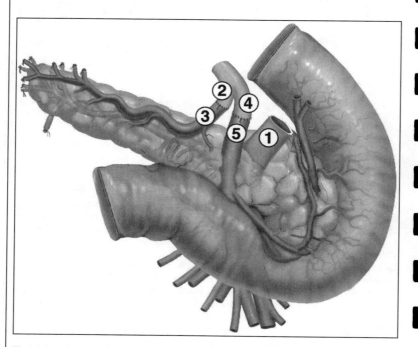

Fig. 3.2. Posterior view of the pancreas on the backtable. The portal vein may need some further dissection to free up approximately 2 cm of its length (1). Finally, the arterial anatomy is reconstructed by performing a Y graft using the donor iliac arteries in which the hypogastric artery (2) is sewn to the splenic artery (3) and the donor external iliac artery (4) to the superior mesenteric artery (5), both end-to-end. The details of the end-to-end anastomosis suture technique is shown in Figure 3.3.

four knots and one of these two tied sutures is then sewn the 25% circumference to the next corner stitch. This corner is also tied in four knots, and one suture is tied in seven knots to the just finished running suture. These steps are shown in Figure 3.3. No sutures are cut until all four corner segments of the anastomosis have been finished; this is to keep the anastomosis stable and open at all times during the entire suturing of the anastomosis.

The external iliac artery from the donor is now sutured to the SMA in the exact same fashion. It should be noted that the internal as well as the external iliac

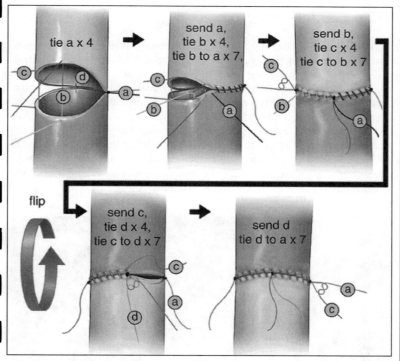

Fig. 3.3. The suturing details of the four-corner end-to-end arterial anastomosis. Two of these constitute the Y graft between the donor iliac arteries and the splenic and mesenteric arteries, respectively (Figs. 3.2 and 3.4). The Gore-Tex CV6 (needle TT9) suture is used (Table 1.4 and 1.5) and placed in the four corners. Each suture is attached to a rubber shod hanging over the edge of the shallow bowl. This keeps the anastomosis lumen open and creates a stable operating field. The most convenient corner stitch (a) is tied in four square knots. One of these is attached to a rubber shod and again placed under slight tension, hanging off the edge of the flat bowl. The second suture is now run to next corner stitch (b), which also is tied in four knots; one of these is tied in seven knots to the suture (a) that has arrived from the first corner. Seven knots in this final tie are needed with the Gore-Tex suture (Table 1.4). The seven knot suture, even though final, is not cut but attached to a rubber shod to keep the anastomosis in progress stable. The remaining stitch at this second corner (c) is now sent to the third corner, and the tying steps repeated until the four corners are completed with suture (d) tied in seven knots to the remaining suture at the starting corner (a).

arteries are cut quite short (Figs. 3.2 and 3.8). Posterior view of the pancreas with the completed Y graft, as it appears on the backtable, is shown in Figure 3.4. If the pancreas has been procured by another team or has been separated on the back-table in an en bloc liver/pancreas removal package, careful attention must be given to the vessels that have been cut. In the author's experience, marked severe bleeding may occur at reperfusion if meticulous tying of multiple cut vessels has not been done during the procurement dissection or at the back-table work.

THE PANCREAS IMPLANTATION

While in 84% of cases in the US the pancreas transplant is done simultaneously with a cadaver kidney transplant (Table 1.1, chapter 1), this chapter will only describe the pancreas organ implant. The pancreas can also be transplanted alone after a previous kidney transplant or before the patient has developed end stage renal disease. The kidney transplant portion of this procedure is identical to that described in chapter 2. The kidney is most commonly placed in the left iliac fossa as part of this intraperitoneal procedure. Some surgeons place the kidney extraperitoneally by creating and dissecting a pocket between the peritoneum and the abdominal wall. This technique may facilitate a safer kidney core needle biopsy. In the author's experience involving more than 20 biopsies under ultrasound guidance in kidneys placed intraperitoneally, no technical complications have occurred.

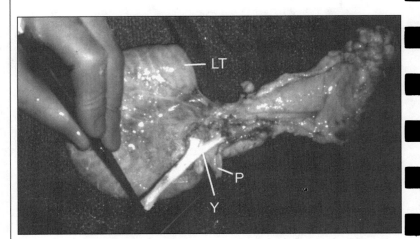

Fig. 3.4. This photograph shows the completed back-table work of the posterior view of the pancreas showing the portal vein (P), the Y graft (Y) consisting of the donor common iliac artery with the external iliac artery anastomosed to the superior mesenteric artery and the hypogastric artery anastomosed to the splenic artery. The spleen has been removed and the pyloric side of the donor duodenum oversewn with a running Prolene® suture. The ligament of Treitz (LT) end of the duodenum has a staple line that will be oversewn and inverted after reperfusion.

The pancreas is usually placed in the right iliac fossa, since the right iliac vessels are slightly more accessible. Also, the authors prefer to use the left kidney in case of simultaneous kidney/pancreas transplant because of the longer vein and the shorter operative time needed to prepare the left kidney. The right kidney is, of course, perfectly fine as well. The pancreas is transplanted first. Others may prefer to implant the kidney before the pancreas.

3

Skin Incision

The authors use a midline incision from the pubic bone to 5 cm above the umbilicus (Fig. 3.5). If the incision is extended too far, it is more difficult to keep the small bowel inside the abdomen and may, consequently, not improve exposure. Prior to surgery the bladder should be expanded with saline containing an antibiotic (i.e., cefalexin 500 mg in 500 cc 0.9% NaCl) with 200-300 ml usually infused easily in the often enlarged diabetic bladder. The round ligaments in females are ligated and divided. A vessel loop is placed around the spermatic cord in males. The entire bladder is mobilized on each side, as well as the peritoneal reflection between the bladder and both lateral pelvic walls. The authors also prepare the midline cystostomy down to the bladder mucosa at this point. The Foley clamp is removed to empty the bladder, which markedly improves exposure and facilitates the often deep placement of the pancreatic portal vein anastomosis.

As with the kidney transplant alone (chapter 2) the authors use the Bookwalter retractor (Codman, Raynham, MA). Other retractors are commercially available which also function well in this setting, depending on the surgeon's personal preference. The arrangement of retractors for the Bookwalter retractor is shown in Figure 3.6. By placing the Bookwalter ring first over the left side and exposing and mobilizing the left iliac arteries and veins for the kidney implant, the retractors need to be realigned only twice, since the pancreas ideally is transplanted on the right side early in this procedure.

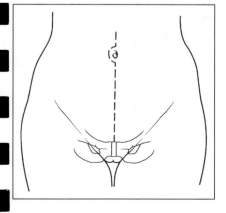

Fig. 3.5. The combined kidney pancreas transplant is an intra-abdominal procedure and is performed through a midline incision from the pubic bone to approximately 5 cm above the umbilicus. This incision is the same as that used for a third intra-abdominal kidney transplant (see chapter 2, Fig. 2.10, incision c).

MOBILIZATION FOR PANCREAS IMPLANTATION

Two malleable retractors are placed, one to retract the small bowel towards the upper abdomen, and a second to hold the mobilized, empty bladder and the sigmoid colon to the left (Fig. 3.6). The cecum is usually mobilized by incising the peritoneum and also 5-6 cm along the ileum, which facilitates placement of the tail of the pancreas in the right flank. The cecum mobilization may not be necessary, depending on the anatomy of the individual recipient.

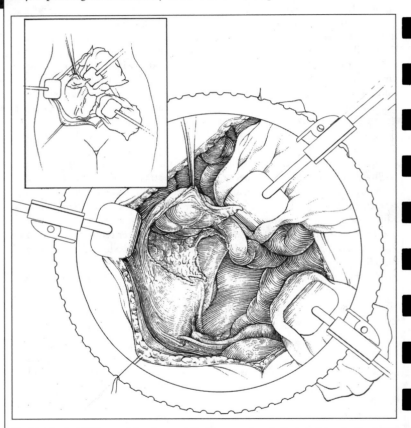

Fig. 3.6. After the left iliac vessels have been mobilized in a procedure similar to that described in chapter 2 for kidney transplant alone, the Bookwalter retractor is moved and arranged from left to the right side. Two wide malleable blades are used to retract the bladder and the sigmoid colon towards the left lower quadrant and the small bowel proximally towards the left upper quadrant. Moist OR laps behind the retractor protect the visceral structures and help keep the small bowel out of the field behind the retractor. A right angle retractor may be used to retract the mobilized cecum to the right upper quadrant (not shown). One or two zero silk sutures may further improve the exposure, with one distal in the direction of the iliac vessels, and finally one in the right lateral flank over the iliac crest (a retractor may also be used, as shown) tied to the Bookwalter ring.

The authors do not ligate lymph vessels when intra-abdominal transplants are performed, since lymphocele formation is very unlikely. In the authors' experience, only one such case has occurred in 39 kidney/pancreas transplants. However, one may encounter temporarily increased abdominal fluid collection after a simultaneous kidney/pancreas transplant as a result of leaking lymph vessels.

This mobilization technique of the right iliac artery and vein is similar to that described for the kidney transplant (chapter 2). Some surgeons routinely ligate and divide the hypogastric artery and vein to improve the exposure and facilitate the vascular anastomoses. In the author's experience, this may be necessary in only about 10-15% of the cases (Fig. 3.7). Some transplant surgeons prefer to make a donor external iliac vein interposition elongation graft on the portal vein. The authors do not think that this is necessary; it fact, it may contribute to venous thrombosis formation or other technical problems. In contrast, a shorter portal

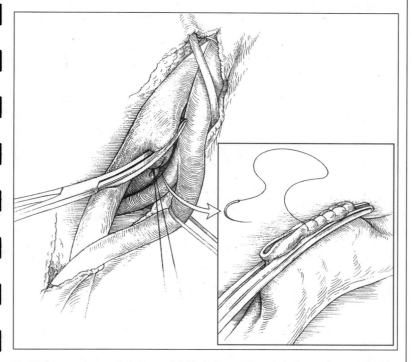

Fig. 3.7. Some surgeons routinely ligate and divide the internal iliac vein for increased exposure, making the portal vein anastomosis easier. This requires careful dissection to avoid severe, hard to control deep venous bleeding. Likewise, some surgeons ligate the hypogastric artery for improved exposure. Different techniques for the hypogastric vein division may be employed, depending on local anatomy, varying from simple suture ligature to division between vascular clamps with a running suture on the hypogastric vein towards the common iliac vessel (insert). In the authors' experience this procedure is not often required.

3

vein will fit better when the pancreas is finally set down in the iliac fossa. The overall pancreas and vascular anastomosis anatomical positions are shown in Figure 3.8.

THE PORTAL VEIN ANASTOMOSIS

The detailed technique for suturing the portal vein to the external iliac vein is described in Figure 3.9. It has been the authors' preference that after the corner stitches have been placed and tied, a retention suture is placed through the back wall to pull it through the front wall to facilitate suturing the back wall from the inside (from the patient's right side). This technique could be used in any vascular anastomosis as dictated by local circumstances and anatomy. In this particular situation, it is usually easier than trying to sew the medial side suture line from the outside. Once the back wall is completed, the retention suture is cut and pulled. The front wall is then finished, using a running suture technique identical to that described for renal vein anastomosis (chapter2, Fig. 2.14). The author prefers the Gore-Tex® suture CV6 on the TT13 needle. The smaller TT9 needle may be preferable with less bleeding from needle holes, but this may present a technically challenging situation. The authors leave the vascular clamps on the external iliac vein while completing the arterial anastomosis.

THE ARTERIAL ANASTOMOSIS

This anastomosis is an end donor common iliac artery to side recipient external iliac artery. Only a short segment of the donor Y graft is used (Fig. 3.8 insert b). As was the case with the vein, the authors prefer velcro clamps on the external iliac artery. Usually it is sufficient to mobilize the external iliac artery up to the level of the hypogastric artery and place the clamp accordingly. Depending on local anatomy and exposure, the technique for sewing the artery can be similar to the technique described for the portal vein (Fig. 3.9) or identical to that for the renal artery (chapter 2, Fig. 2.15). The difference is that the authors place a distal corner stitch, which is tied, while the proximal corner stitch is left open under constant tension to improve exposure during suturing the first, usually lateral (right) side of the anastomosis. The back wall (left side) can either be sewn from inside, as was the case with the portal vein, or from the contralateral side, in which case no back (left) suture line retention suture is to be placed. The artery is usually placed in an approximately 60° angle to the external iliac artery (Fig. 3.8). Only a small portion, or approximately 1-2 cm of the common iliac artery from the donor, is used (Fig. 3.8). As was the case for the portal vein anastomosis, the authors prefer the Gore Tex® suture CV6 on the TT13 needle. This larger needle is optimal for easy suturing. After completion of the anastomosis, the authors place Heifetz clamps or velcro bulldogs on the Y graft before releasing the distal external iliac artery velcro clamp. If there is no significant bleeding, then the proximal arterial velcro clamp is removed. A velcro clamp is now placed on the portal vein. Next, the proximal velcro clamp on the external iliac vein is removed. If there is no bleeding, the distal iliac clamp is removed. The order in which the clamps are removed

3

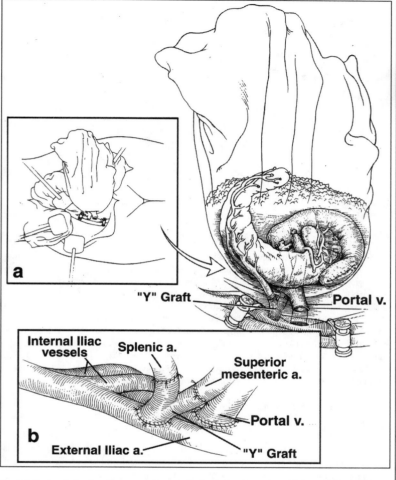

a

"Y" Graft   Portal v.

Internal Iliac vessels   **Splenic a.**

Superior mesenteric a.

Portal v.

b

External Iliac a.   "Y" Graft

Fig. 3.8. During the implant the pancreas is covered in ice slush and rests in an OR lap with the portal vein and the arterial Y graft hanging down through a small hole in the gauze (insert a). The detailed anatomic arrangements of the two vascular anastomoses are outlined in (insert b). The angles and locations of these two anastomizes are important, both for prevention of thrombosis as well as the immediate technical aspects of performing these two often difficult and deep anastomoses. The portal vein anastomosis technique is detailed in Figure 3.9.

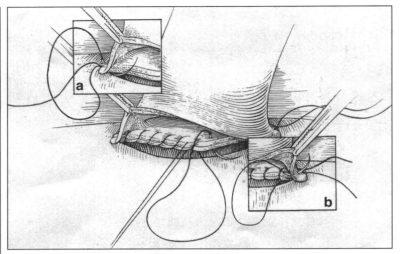

Fig. 3.9. Detailed suturing technique of the end portal vein (PV) to side external iliac vein (EIV) anastomosis. This anastomosis is very similar to that for the renal vein to EIV (chapter 2, Fig. 2.14), but is more challenging because of the shorter portal vein (PV). The velcro vascular clamps are applied (Fig. 3.8). The venotomy is made with a #11 blade and extended with Diethrich-Potts scissors. Corner stitches are placed as shown in Figure 3.8 and tied with four square knots. The back wall (left aspect of the anastomosis) is more easily sutured from the inside (right side). This is facilitated by using a retention suture (6-0 Prolene® on a CV1 needle) on the midportion of the back wall. The first stitch goes outside-in on the proximal corner of the EIV (insert a) and then the back wall is run to the proximal corner. This assumes that the surgeon is right-handed and is standing on the patient's right side, and that the pancreas is transplanted into the right iliac fossa. The last stitch goes inside-out on the portal vein and is tied to one of the previously tied corner stitches (insert b). The retention suture is cut and removed. The front wall is run all the way from one corner to the other or, alternatively, the two sutures meet midway and are tied in seven square knots, as was described for the renal vein anastomosis.

is important. By removing the proximal venous clamp first, only slight pressure is applied to the portal vein anastomosis. This allows occasional large leaks or missed branches to be repaired or ligated. Removal of the distal clamp first, while the proximal clamp is still on, may potentially rupture the anastomosis from the high pressure built up in the leg. The temporary clamps on the artery and portal vein are removed to let the pancreas reperfuse. Usually, there are a few significant bleeders on the pancreas needing immediate attention. Hemoclips, mosquito hemostats and sutures are used depending on location, access and the surgeon's preference. Most oozing will stop within a few minutes after letting the pancreas rest, packed in an OR lap. When complete hemostasis is obtained, the exocrine drainage procedure follows.

EXOCRINE DRAINAGE

BLADDER DRAINAGE

According to the UNOS registry, in the United States, the urinary bladder is currently used in 96% of cases as the route for exocrine secretion drainage from the transplanted pancreas. In the simultaneous kidney/pancreas transplant procedure, the bladder is, therefore, used for both the exocrine and ureteral drainage. There are several different techniques described to accomplish this. Perhaps most surgeons use a two-layer handsewn donor duodenum to dome of bladder cystostomy. An absorbable suture (4-0 PDS-II® , SH needle or 4-0 vicryl® on a PS-2 needle) is used for the inner mucosal layer followed by interrupted permanent 3-0 suture between muscle layers.

As an alternative, the authors have used a circular stapling device (chapter 1, Table 1.6). The excess distal portion of the mobilized duodenum is used as the stapler insertion site. The duodenum is opened next to the staple line and first rinsed with saline solution (Fig. 3.10a). The cap of the stapler is removed and the sharp leading spike is extended maximally. (For proper use of the specific circular stapler used, the reader should study the package insert in detail.) The stapler is inserted to the midportion of the donor duodenum, which is then perforated. A site is identified on the dome of the bladder and again, the spike is used to penetrate the entire bladder wall. Inside the bladder the cap is reattached, the stapler device is screwed into position as indicated on the stapler, and the instrument fired (Fig. 3.10b). The cap and the stapler are next removed from the duodenum.

The excess duodenum is now removed, using a TLC or TCR 55 stapler (Table 1.6, chapter 1) and is also oversewn with a 4-0 Prolene® suture, using the same stapler line inverting technique described for the pylorus side on Fig. 3.1, step 4. The cystostomy staple line is oversewn with a running 5-0 PDS-II® from inside the bladder, mainly for hemostatic reasons. During the kidney implant, the bladder midline cystostomy is left open and packed with gauze to absorb the pancreatic secretions and prevent it from leaking into the abdominal cavity. After the kidney transplant and ureter implant, the cystostomy is closed using 4-0 PDS-II® sutures (on an SH needle) in three layers, as described in chapter 2, p 47 for the kidney transplant procedure. In the authors' experience with 22 consecutive duodenocystostomy stapled pancreata, there have been no anastomotic leaks. There were one early (6 weeks) and two late (12 months) leaks from the duodenum staple line area.

ENTERIC DRAINAGE

Initial poor results in the early 1970s from various enteric drainage modalities have tainted the small bowel drainage procedure with a bad reputation associated with high morbidity, mainly from anastomotic leaks with abscess formation and sepsis, as well as mortality. Bladder drainage was also introduced to enable urinary amylase to be used to diagnose pancreatic rejection. There are several enteric

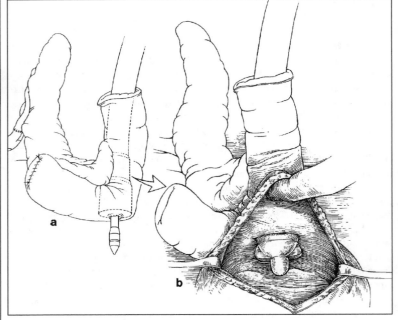

Fig. 3.10. The donor duodenum side to recipient small intestine to dome of bladder stapled anastomosis. A circular stapler (usually CDH #25) is inserted from the ligament of Treitz end of the donor duodenum to the lateral aspect of the duodenum, where the sharp spike will exit through the duodenum (Fig. 3.10a) and then through the dome of the bladder (Fig. 3.10b). The cap is attached inside the bladder onto the instrument and is tightened as per manufacturers' insert instructions, and fired. The cap is removed and the stapler retracted through the duodenum. The suture line inside the bladder is oversewn with a running 4-0 PDS-II® for hemostasis.

drainage modalities regarding bowel arrangement (i.e., Roux-en-Y) and the level of anastomosis site of the recipient small bowel. Furthermore, the anastomoses can be stapled or hand sewn. As more centers are using enteric drainage, the next several years will more clearly determine what methods render the best outcome. The authors' last 17 consecutive enteric drainage procedures compare favorably with our first 22 bladder drained pancreata (Table 3.3 and chapter 1, Table 1.7). The following outlines our slightly biased view of enteric drainage, namely the stapled side donor duodenum to side recipient enteric anastomosis.

The Side Donor Duodenum to Side Recipient Enteric Stapled Anastomosis

After complete hemostasis is obtained, the pancreas is placed to rest lateral to or even behind the mobilized cecum. Any portion of the small bowel that reaches without tension may be used for the anastomosis. Often, a site 20-30 cm from the cecum is chosen. The selected loop of small bowel (usually the ileum) is lined up parallel to and on top of the donor duodenum. Two silk holding sutures are placed

(Fig. 3.11a). Using electrocautery, the serosal layers are opened at the stapler insertion site and the enterotomies appropriately widened with a hemostat. The donor duodenal contents are emptied with an intestinal suction device and irrigated clean with saline. The subcutaneous administration of 100 mcg Somatostatin® at the beginning of surgery prevents the donor duodenum from being extended by large amounts of pancreatic secretion (for pre- and intraoperative orders, see chapter 6, Tables 6.3-6.5). The staple device (TLC or TCR 55) is now inserted (Fig. 3.11b). The small bowel is properly aligned and the instrument is fired. The staple line is inspected for bleeding inside the bowel (Fig. 3.11 insert c). The staple defect connecting the recipient's small bowel and the donor duodenum is now closed transversely in two layers, using a running 4-0 absorbable suture (i.e., Vicryl®, 5-0) for the mucosa layer and a 3-0 interrupted silk for the serosal layer (Fig. 3.11c).

SURGICAL COMPLICATIONS AFTER PANCREAS TRANSPLANTATION

BLADDER DRAINAGE

Bladder drained pancreas transplants are plagued with urological and metabolic complications (Table 3.3 and chapter 1, Table 1.7) requiring enteric conversion in 15-20% of most reported series.

- Hematuria is almost universal and requires intervention in about 30% of cases, such as Foley catheter placement, irrigation and/or cystoscopy for evacuation of clots.
- Urinary tract infections occur in up to half of all cases, and are likely induced by the irritating effect of the exocrine secretion on the bladder mucosa.
- The burning pain from urethritis, especially in males, is a severe complication leading to urethral strictures. This requires urological expertise in addition to enteric conversion.
- Bladder stone formation may occur on the staple or suture line and may be a cause of hematuria. The diagnosis is confirmed and stone removal is performed during a cystoscopy procedure. In the authors'

Table 3.3. Metabolic consequences of pancreatic transplantation: bicarbonate (meq/day) profile in 39 patients, 22 of whom had exocrine drainage to the bladder and 17 to the small intestine

Bicarbonate	Bladder Drained	Enteric Drained
Current Dose (meq/day)	121.0 + 97.3	0
Dose Range (meq/day)	72–360	0
Number Tablets/day*	9–45	0

* 1 Tablet HCO_3^- = 9 meq of HCO_3^-

3

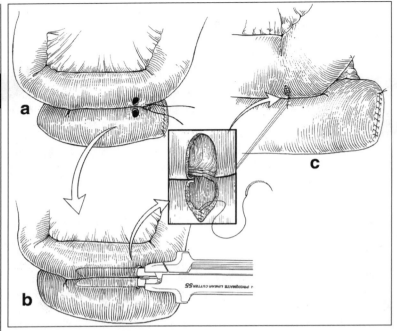

Fig. 3.11. The side donor duodenum to side recipient small bowel stapled anastomosis. In the authors' opinion, any segment of the small bowel that reaches to the donor duodenum can be used. The small bowel segment is attached to the duodenum through two silk sutures (Fig. 3.11a). Enterotomies are made in the small bowel and the donor duodenum using electrocautery and a mosquito hemostat. The donor duodenum is irrigated with saline. The TLC or GIA 55 staple device is inserted and fired (Fig. 3.11b). The staple line is checked for completeness and hemostasis (Fig. 3.11 insert c). The defect between the small bowel and the donor duodenum is closed with a running 4-0 Vicryl® in the mucosa and then interrupted 3-0 silk serosal layer (Fig. 3.11c). If not done before, 100 mcg of Somatostatin® should be given subcutaneously, to avoid intestinal distention (chapter 6, Tables 6.4 and 6.5).

experience, two small (2-3 mm) anastomotic stone formations occurred with hematuria in 22 bladder drained pancreata.
- Metabolic complications include acidosis and dehydration from lack of adequate salt (bicarbonate) and fluid intake. In our series, 21 of 22 bladder drained patients (98%) have been readmitted within the first year for correction of severe acidosis with or without dehydration, suggested by increased serum creatinine which required intravenous fluid and bicarbonate administration. The magnitude of pancreatic bicarbonate excretion is suggested in Table 3.3, with patients requiring up to 400 meq of bicarbonate daily, corresponding to 45 tablets of bicarbonate (1 tablet = 9 meq of HCO_3).

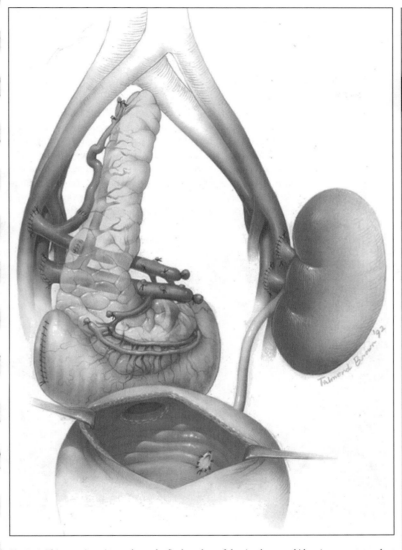

Fig. 3.12. This overview picture shows the final product of the simultaneous kidney/pancreas transplant with bladder drainage.

3

- Early urine leaks are technical in nature and usually require surgical correction in addition to prolonged Foley catheter urinary drainage. Small urine leaks may be hard to diagnose, and therefore, are sometimes treated with a Foley catheter based on a high degree of suspicion related to symptoms. In cases of infection and sepsis from urine leaks leading to abscess formation, removal of the pancreas at an early stage may be wise in order to avoid major morbidity including death from septic complications. Late leaks occur from the staple line sites at the duodenum and are caused by high pressure in the duodenum during urination. These are, again, usually small leaks and they are often hard to diagnose with a cystogram (Fig. 3.13a). The treatment is Foley catheter placement for six weeks, or enteric conversion, depending on the patient's condition, degree of leak and the surgeon's personal preference. A penile prosthesis reservoir (Fig. 3.13b) may, for the untrained eye, mimic a urine leak. A normal cystogram 10 days after surgery is shown in Figure 3.13c.

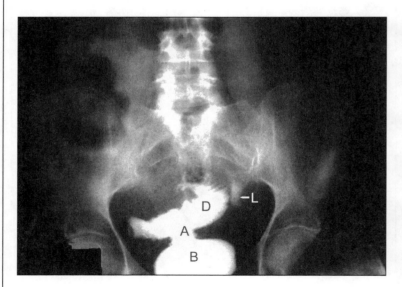

Fig. 3.13a.

Fig. 3.13. A cystogram one year after pancreas transplant shows a small urine leak (L) at the ligament of Treitz staple line (Fig. 3.13a). A contrast filled penile prosthesis (P) reservoir placed between the bladder (B) and the rectus muscle may, to the untrained eye, resemble a urine leak (Fig. 3.13b (opposite page)). A normal cystogram taken 10 days after transplantation before removal of the Foley (F) catheter is depicted in Figure 3.13c (opposite page). The donor duodenum (D) is also filled with contrast. See also chapter 9, Figures 9.4-9.6.

3

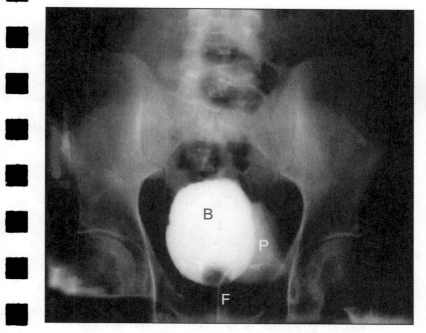

Fig. 3.13b.

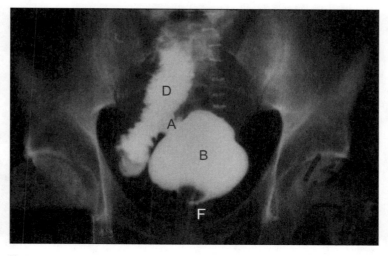

Fig. 3.13c.

ENTERIC DRAINAGE

A contrast enhanced CT scan is a valuable diagnostic tool (Fig. 3.14). Even though the authors have not experienced anastomotic leaks in the last 17 consecutive side-to-side donor duodenum to recipient small bowel exocrine drained pancreas transplants, this is the most dreaded complication. Optimal surgical technique during organ procurement including donor management and during the implantation procedure coupled with sound clinical judgment guided by detailed protocols, yields acceptable results in this very challenging patient population.

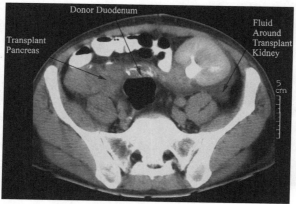

Fig. 3.14a. Contrast enhanced abdominal CT scan showing the enteric drained transplanted pancreas in the right and the kidney in the left iliac fossae.

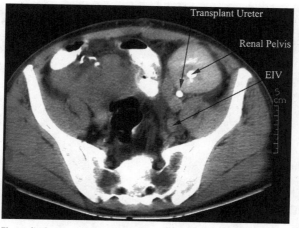

Fig. 3.14b. The pancreas appears enlarged (swollen) supported by clinical symptoms and elevated amylase concentrations. The ureter and a lower pole calyx appear in Figure 3.14b.

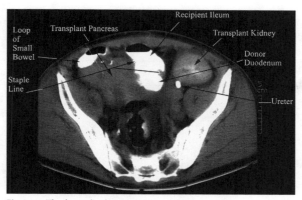

Fig. 3.14c. The donor duodeno-enteric anastomosis is evident.

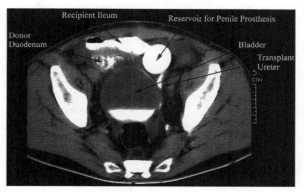

Fig. 3.14d. A penile prosthesis reservoir is located between the filled bladder and the abdominal wall. See also Figure 3.13b.

SURGICAL AND VASCULAR COMPLICATIONS

Major surgical complications after pancreas transplantation in our experience have occurred in 10% (4/39) of patients (Table 3.4a). The main cause of pancreas losses is vascular thrombosis, which has also occurred in 10% (4/39) of patients (Table 3.4b). The low rejection rate in this series likely reflects the aggressive, early immunosuppression in our program (See chapter 6). There seems to be no difference in the incidence of nonurological complications between bladder and intestinal drainage.

ABDOMINAL CLOSURE

The midline incision is closed using interrupted 0-0 Prolene® sutures or 0-0 Nurolon® (Control Release® CT-1 needle). Any absorbable suture material is advised against, since the wound healing process may take many weeks to months. The author has, in fact, seen one dehiscence at 12 weeks posttransplantation and

Table 3.4. *Major surgical complications (a), and pancreas graft losses (b), in the authors' series of 39 simultaneous pancreas/kidney transplants, 22 of which had bladder and 17 had enteric exocrine drainage.*

a.

	Bladder drainage n=22	Enteric drainage n=17
Wound infection	0	0
Abscess	1	0
Postoperative bleeding	0	1
Wound dehiscence	1	0
Pancreatitis	1	0
Total	3/22 (13.6%)	1/17 (5.9%)

b.

	Bladder drainage n=22	Enteric drainage n=17
Death	1	1
Vascular thrombosis	3	1
Rejection	1	0
Abscess	1	0
Total	6/22 (27.3%)	2/17 (11.8%)

two incisional hernias in cases that were closed with running 0-0 PDS-II® suture. As in the case with cadaver kidney transplantation, the author strongly advises against using any drains in pancreas transplantation. Meticulous, atraumatic, exact surgical technique and hemostasis are key components to success. Also, by using restricted electrolyte fluid infusion during the procedure with more colloid administration, the patient will experience less intestinal edema and an earlier return of bowel function (See Table 6.4, chapter 6).

SELECTED READING

1 Davidson I, Lu C, Melone D et al. Cyclosporine therapy in pancreas transplantation. Trans Proc 1996; 28(4):2134-35.
2 Davidson I, Lu C, Munschauer C et al. Intestinal drainage of the exocrine pancreatic allograft revisited. A better choice? Trans Proc 1996; 29:639.
3 Dawidson IJA, Rossi P, Meyer D. Multiple organ procurement protocol: details and variations of surgical technique. In: McClelland RN, ed. Selected Readings in General Surgery. Dallas: The University of Texas Southwestern Medical Center at Dallas Publishing, 1993:20(7).
4 Dawidson IJA. Current state of pancreatic transplantation. In: Current Topics in General Surgery. Dallas: University of Texas Southwestern Medical Center at Dallas Publishing, 1995:53-59.
5 Ozaki CF, Stratta RJ, Taylor RJ et al. Surgical complications in solitary pancreas and combined pancreas-kidney transplantation. Am J Surg 1992; 164:546-551.
6 Sollinger HW, Knechtle SJ, Reed A et al Experience with 100 consecutive simultaneous kidney-pancreas transplants with bladder drainage. Ann Surg 1991; 214(6):703-711.

Laparoscopic-Assisted Living Related Nephrectomy

4

Michael Edye, Robert Waterhouse

INTRODUCTION

The evolution of advanced laparoscopic techniques now permits an operative team with adequate experience to perform donor nephrectomy. At present, this operation is performed intraperitoneally, and the kidney extracted using a retrieval bag inserted through a 6-7 cm long incision sited for minimal impairment of function and cosmesis. Avoidance of a lengthy loin incision with possible rib resection make this operation an attractive alternative for potential donors. With the prospect of a two-day hospitalization and rapid return to normal activity, the donor pool is likely to be expanded by healthy individuals needing to return to work as promptly as possible who would otherwise not have considered kidney donation.

PERSONNEL AND EQUIPMENT

An ideal team for this procedure is a urologist with experience in advanced laparoscopic techniques, or a urologist working with a dedicated laparoscopic surgeon. Complete familiarity with the surgical anatomy from the laparoscopic point of view is essential. Gentle, meticulous surgical technique minimizing hemorrhage is necessary to avoid a crush or puncture injury to the vessels, ureter and kidney parenchyma.

A major asset is nursing support staff familiar with both donor nephrectomy and advanced laparoscopic techniques. Operating room nurses should spend time in the animal laboratory with the surgical team practicing the sequence of steps necessary during the procedure. Concentration, good preparation and full teamwork are essential prerequisites for this complex operation.

A good quality high resolution video camera with high intensity light source gives clear operating conditions. Three chip or the latest generation electronically enhanced single chip cameras are ideal. The Olympus OTVS-5 system is excellent but other video combinations are equally good.

For donors with thin to moderate thickness of their abdominal wall, a 5 mm 45° telescope provides an excellent image while minimizing the size of the abdominal puncture. On the left three 5 mm ports and one 10 mm or 12 mm port are necessary for mobilization of the kidney. A 15 mm cannula inserted through the extraction incision for the retrieval device is necessary towards the end of the case. Good quality insulated grasping instruments of the fenestrated, atraumatic type are useful for handling soft delicate structures, but a narrow pointed grasper will be needed for accurate dissection around the vessels. A right angle dissector is also necessary for vascular dissection and passing loops around the vein and artery. Colored silicone vessel loops are used to tag artery, vein and ureter and allow gentle retraction of these structures.

The laparoscopic coagulating shears (LCS, Ethicon, Somerville, NJ) are particularly valuable in the dissection of vascular fat, which they cut and cauterize exceedingly well. Veins of 3-4 mm diameter can be coagulated and divided with a single application of this instrument. Depending on the setting, arteries of 2-3 mm diameter can also be treated this way. The disposable hand piece is expensive, somewhat cumbersome, and requires a 10 mm cannula but with proper use is a superb aide to dissection at many stages of this operation.

A single multiple clip applier (20 clips) should be enough for the case.

Large vessels can be ligated and divided in a number of ways depending on their size and surgeon's preference. The easiest but most expensive technique is to use an endoscopic stapler which will require a 12 mm cannula for insertion. If the artery is less than 5 mm in diameter, it can be divided after application of 2 clips. Suture ligation using 2-0 Ethibond® or 3-0 Gore-Tex® and an extracorporeally tied slip knot is a good method for ligation of the vein as it preserves length and allows the kidney to empty of blood if divided without ligating the kidney side of the vein.

For right nephrectomy a laparoscopic Satinsky clamp (Microfrance, Bourbon-l' Archambault, France) is necessary. Because of the depth of the jaws, this will only pass through a 12 mm cannula.

Straight endoscopic scissors allow a cleaner cut on the artery and vein, but curved disposable endoscopic shears are more useful during general dissection.

PREOPERATIVE WORK-UP

Because the left renal vein is longer, the left kidney is the preferred donor side. Recent experience with right donor nephrectomy has proved this to be a feasible alternative if the anatomy on the left is unsuitable. Although donor selection is beyond the scope of this chapter, preoperative evaluation must establish that renal function is normal and symmetric. We currently use renal angiography to determine the vascular pattern. Three-dimensional reconstruction of a spiral CT may provide equivalent information, although the total load of contrast is much the same. Double gadolinium magnetic resonance scans (for both artery and vein imaging) may prove comparable. What matters is that the chosen imaging modality can reliably demonstrate multiple renal arteries, arterial occlusive disease, venous abnormalities such as circumaortic or retroaortic left renal veins, and ureteral anomalies (Figs. 4.1a and b).

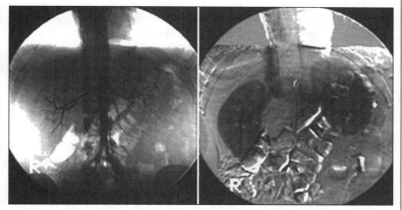

Fig. 4.1a. Normal angiogram showing a single left renal artery, normal vein, and a single ureter.

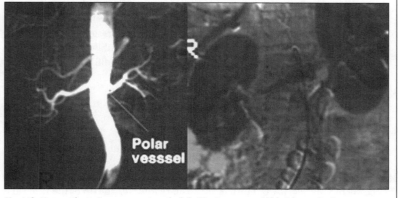

Fig. 4.1b. Two renal arteries are present on the left side. At surgery a bifid right renal vein was encountered, necessitating anastomosis of both veins in the recipient.

A checklist with the results of compatibility testing and other studies should be kept so that the file is complete before the donor is submitted to surgery. This is especially important in a busy transplant center with changing personnel. One person should be designated as responsible to ensure that the results of all compatibility testing are available and satisfactory before the donor is taken to the operating room.

Measures to prevent thromboembolic complications are advisable. Currently, our donors receive prophylactic subcutaneous heparin (5,000 units every 12 h) and calf compression boots are used intraoperatively.

SURGICAL TECHNIQUE

The principles of this procedure are the same as for open donor nephrectomy but with a minimally invasive surgical slant. There can be no compromise in the essential steps of the operation.

Mobilization of the kidney within Gerota's fascia is deferred until all anterior vascular dissection is complete. Otherwise the kidney will fall into the operative field, making subsequent medial dissection more difficult.

Dissection of the ureter must be atraumatic and must preserve the blood supply emanating from the renal pelvis and traveling along the ureter. Dissection of the vein and artery must be perfect without rough handling that could cause arterial spasm, full thickness or intimal injury to either vessel.

The kidney parenchyma must not be penetrated by any of the narrow laparoscopic instruments. The renal capsule should be kept intact. Warm ischemia should be minimized. It should be possible to clamp and divide the vessels and extract the kidney in less than four minutes.

DONOR POSITION

While the patient is supine after induction of general anesthesia and endotracheal intubation, a urinary catheter is inserted for the duration of the case. The skin crease and midline or Pfannenstiel incision or infraumbilical extraction incision are marked with an indelible pen as the correct site is difficult to judge when the donor is positioned.

LEFT NEPHRECTOMY

The donor is rolled into the standard nephrectomy right lateral decubitus position so that if conversion to an open procedure is necessary there will be adequate distraction of the costal margin from the iliac crest, as is achieved with the bending of the table, use of bolsters and axillary roll, etc. The tip of the 12th rib is marked and a line carried forward to the umbilicus. The point of intersection of this line with the lateral border of the rectus is a good position for a 10 or 12 mm cannula (#2) for the clip applier, LCS or stapler used in the right hand of the surgeon. Two other ports are sited along the lateral border of the rectus. Eight cm above the umbilicus is sited the cannula for the laparoscope (#1) and a further 8 to 10 cm above that a 5 mm trocar for the surgeon's left hand (#3); 10 or 12 cm laterally on the line marked is a fourth puncture for a 5 mm cannula (#4) for suction and retraction (Fig. 4.2).

4

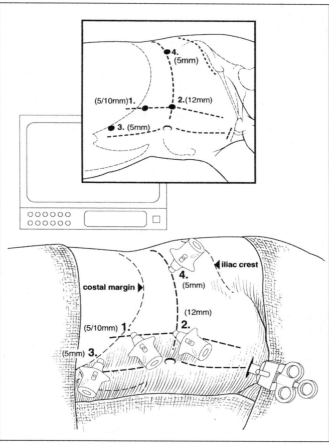

Fig. 4.2. The patient is in the standard right lateral decubitus position for nephrectomy. The laparoscope port is positioned 8 cm above and lateral to the umbilicus (#1). The main operating port (#2–10 or 12 mm) is on the lateral border of the rectus intersecting with a line drawn from the 12th rib, approximately 8 cm caudad to port #1 (roughly level with the umbilicus). Port #3 is 8 cm caudad to port #1 and is used by the surgeon's left hand. Port #4 is for retraction, and suction is approximately 12 cm lateral to port #2 on the line marked. Port #5 (15 mm) is placed in the midline through a short Pfannenstiel incision.

Pneumoperitoneum is induced by Veress needle puncture directly through the abdominal wall at the costal margin in the midclavicular line. The laparoscope cannula (#1) is inserted first. By rotating the 45° telescope to look upwards at the abdominal wall the three subsequent cannulae are inserted under vision. The most comfortable position for the operator is to stand facing the donor's abdomen viewing the monitor positioned behind the donor's back. The dissecting instruments are introduced on either side of the laparoscope (#2 and #3), which is controlled by an assistant standing directly behind or to the right of the surgeon. A second assistant standing behind the donor's back can retract or suck through the lateral-most 5 mm trocar (#4).

The splenic flexure is frequently adherent to the abdominal wall and this is divided by sharp dissection using an LCS or scissors. Neither the splenic parenchyma nor colonic wall are handled. Rather, the peritoneal attachments to them are picked up for traction during the subsequent mobilization. The peritoneum of the splenorenal ligament is opened sharply and the cut carried down just lateral to the colon. A thin flap of the peritoneum reflecting from the colon wall is incised so that the colon can be reflected medially. Blunt dissection can be used as long as bleeding is avoided. This plane is essentially avascular. The colon is reflected generously toward the midline and this line of incision is carried superiorly enabling the spleen to be reflected medially out of the operative field. The bulge of the kidney within Gerota's fascia is easily seen below the lower pole of the spleen and behind the colon. A completely avascular plane will be developed as these two structures are reflected medially. It is common to see the inferior mesenteric vein towards the limit of this dissection folding forward in the descending mesocolon (Fig. 4.3). If Gerota's fascia is thin, it may be possible to find the ureter easily as it crosses the iliac vessels. A generous ribbon of fat is left around it, and it is looped with a 7 cm length of yellow vessel loop secured with a clip (Fig. 4.4).

The renal vein can often be seen as a darkening in the fat of the medial part of Gerota's fascia (Fig. 4.3). If this is not evident then following the gonadal vein superiorly will lead to the renal vein (Fig. 4.5). Dissection is carried out in the plane of the vein wall and all of the fat and fascia anterior to the vein is cleared. In the absence of previous surgery or inflammation, this is an absolutely bloodless plane and can be dissected with great accuracy. Another technique to locate the ureter is to follow the gonadal vein inferiorly. The ureter will be seen to pass underneath it as it descends from the renal pelvis. The gonadal vein is divided between clips, superomedial to the ureter, and the dissection carried superiorly to its junction with the renal vein. One or more lumbar tributaries may enter the gonadal vein or the confluence of the gonadal and renal veins. With upward traction on the gonadal vein these branches will come into view and can be cleanly dissected with the right angle instrument. The stump of the gonadal vein is used as a handle to clear the lower border of the renal vein. A 2-0 silk tie passed around the lumbar tributary is tied extracorporeally and the knot snugged down on the side that will remain with the kidney. On the donor side clips can be applied.

As a general rule, clips should be avoided near the main renal vessels as these may prevent proper firing of the stapler later in the case.

4

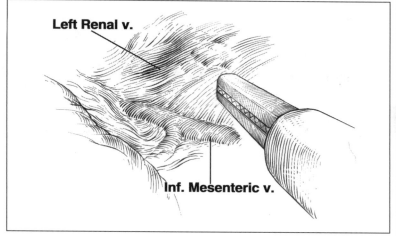

Fig. 4.3. Towards completion of mobilization of the mesentery of the colon, the inferior mesenteric vein is often encountered and can be mistaken for the gonadal vein. Here a darkening visible in Gerota's fascia indicates the site of the renal vein. This will be in the same anatomic plane as the gonadal vein.

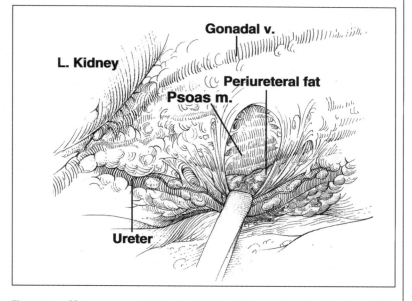

Fig. 4.4. A vessel loop encompasses both ureter and periureteral fat containing blood vessels essential to the ureter. All tissue lateral and posterior to the ureter as deep as the psoas muscle should be included in this sling.

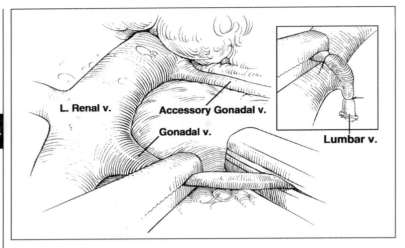

Fig. 4.5. The gonadal vein is elevated and divided well distal to the renal vein and is used as a handle. Final ligature should be tied with silk nearly flush with the renal vein. In this case, because of the small size of these duplicated gonadal veins, division with the LCS was possible. Clips should be limited to the side of the vein to stay.

The adrenal vein will be found on the superior aspect of the renal vein and this is similarly divided. It is important to stay close to the plane of the vein wall as numerous small arterial filaments fill the space lateral to this and will bleed easily if disturbed. Tie on the renal vein side (Fig. 4.6) and clip on the adrenal side and divide the vein between them with scissors. This stump is also a useful handle for dissection of the vein. Returning to the gonadal vein stump, the posterior aspect of the vein is cleared and with care the right angle instrument is passed behind the vein to grasp a short length (about 3 inches) of blue vessel loop (Fig. 4.7) which is secured with a clip. The vein is freed as far medially as possible over the front of the aorta. This dissection is limited by the presence of the superior mesenteric artery medially and the pancreas superiorly. If the spleen and the pancreas seem to be in the way, it is probable that the splenophrenic ligament has not been divided high enough to enable the spleen to fall forward.

Mannitol (12.5 g) is administered intravenously at the commencement of dissection of the vessels to promote renal parenchymal perfusion. The most important technique to prevent vascular compromise is meticulous, gentle, indirect handling of the artery.

Because of the decubitus position of the patient and the laparoscopic view, the artery must be imagined as a structure running vertically behind the vein. If this is not considered, the anterior aspect of the aorta and the celiac ganglion may be dissected in error. The artery is surrounded by fibrofatty tissue, lymphatics and autonomic nerves. All are divided with the LCS. The adventitia of the artery should be sharply dissected on the upper and lower aspect to expose the shiny white arterial

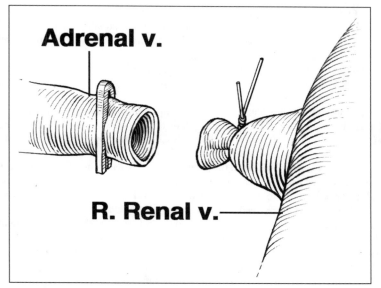

Fig. 4.6. Thread ligature of the adrenal vein is preferred, as clips are more likely to fall off during subsequent dissection and may get in the way should a stapler be used to divide the vein. Clips are acceptable on the adrenal side.

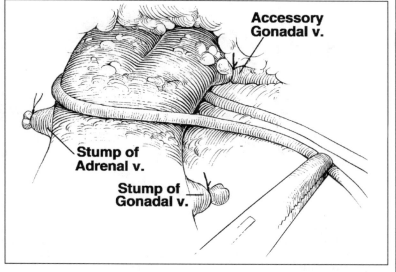

Fig. 4.7. Encircling the left renal vein with a blue vessel loop enables gentle retraction. Excessive traction will stop blood flow in the kidney and adds to warm ischemia.

wall. Only then is it possible to safely pass the right angle instrument behind the artery using gentle side-to-side rocking movements. The jaws should not be spread unless the tips are both visible (Fig. 4.8). The artery is encircled with a white vessel loop secured with a clip.

The areolar tissue proximal to this is gently pushed away with a peanut dissector to the point where the origin of the artery appears to flare on to the aortic wall. No further arterial dissection is necessary.

The adrenal gland is separated from the superomedial aspect of the kidney as far as the upper pole of the kidney. Small direct branches to the gland from the artery must be dealt with individually, but are not large enough to hold clips and can be cauterized or coagulated with the LCS.

The ureter is mobilized from the pelvic brim upwards. By not dissecting in the triangle bounded medially by the ureter and superiorly by the lower pole of the kidney, vessels critical to the ureter will be avoided (Fig. 4.9). The contents of this triangle (ureter, fat) are freed from the psoas muscle posterior to it, creating a largely avascular plane that leads up behind the kidney. At about the point where the ureter passes behind the gonadal vein, Gerota's fascia is cut directly upwards toward the lower pole of the kidney (along the third side of the triangle mentioned above) to expose the capsule. This cut is carried toward the upper pole, intentionally keeping away from the hilar structures. Counter traction is provided

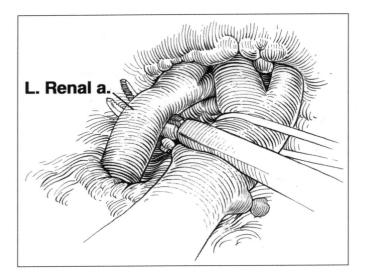

L. Renal a.

Fig. 4.8. Once the upper and lower periarterial adventitia has been cleared from the artery with blunt dissection, the closed tip of the right angle dissector is advanced gently behind the artery with gentle side-to-side movements. Spreading the jaws of the instrument should be scrupulously avoided unless both tips are visible. A white vessel loop gives gentle retraction and is very useful to identify the artery when the kidney has been reflected medially.

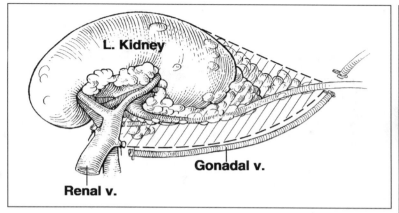

Fig. 4.9. A triangle bounded medially by the ureter, superiorly by the lower pole of the kidney and laterally by a line from the lateral border of the kidney to the point of intersection of the gonadal vein and ureter is a "no-fire zone." Important vessels to the ureter from the renal hilum are present in this tissue.

by an instrument at #4, and after complete clearance of the lower pole of the kidney the instrument at #3 can be used to elevate the lower pole as it is shelled out of Gerota's fascia (Fig. 4.10). This mobilization of the front, back and lateral borders of the kidney proceeds cephalad until the upper pole is free. The kidney will fold further forward. As dissection is carried further medially toward the back of the hilum, the loop on the renal artery will be seen from the posterior aspect (Fig. 4.11).

Circumferential mobilization of the kidney is completed by ensuring that all of Gerota's fascia on the back of the upper pole and between the ureter and psoas muscle are cleared.

COORDINATION WITH RECIPIENT TEAM

Performance of the nephrectomy and transplant by separate teams reduces fatigue and allows for better timing of the recipient procedure. Ideally, once the explant team has the donor kidney on a clean vascular and ureteral pedicle, the recipient is already anesthetized and vessels exposed for immediate transplantation.

FINAL PREPARATION OF THE DONOR KIDNEY

The kidney will be ready for explantation when both vein and artery have been skeletonized, there are no peripheral attachments to the kidney and the ureter is mobilized to the pelvic brim.

4

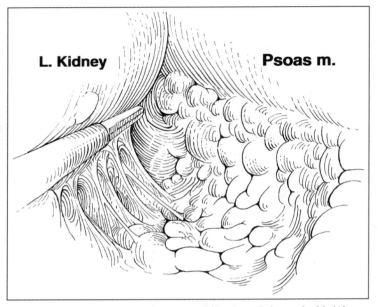

Fig. 4.10. A blunt instrument introduced to port #3 is used to elevate the lower pole of the kidney as it is separated from perirenal fat.

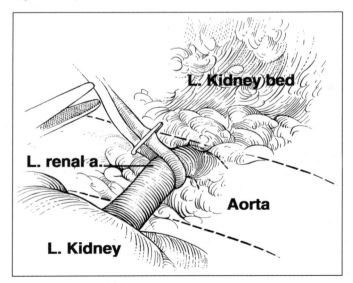

Fig. 4.11. By flipping the kidney to the midline, the posterior aspect of the hilum is exposed. Once all perihilar fat is freed from the psoas muscle behind, the renal artery with its vessel loop will be exposed.

At this stage, a decision regarding ligation of the artery will be made. When in doubt, the vascular stapler is the easiest and quickest option. If so, test positioning of the stapler across the origin of the renal artery is performed.

Provided that the renal vein is divided central to (on the caval side of) the gonadal vein, there will be ample length for implantation. We prefer suture ligation. A 2-0 Ethibond® or 3-0 Gore-Tex® suture is looped around the vein, in at least one position picking up a thin layer of vein wall as for suture ligation (but without transfixation of the anterior and posterior walls of the vein). This prevents the suture ligature from slipping during the subsequent maneuvers. The thread is tied extracorporeally and the knot pushed down until it lies loosely, about 5 mm away from the vein. This thread is brought out through trocar #3 to keep it out of the field during division of the artery.

Twenty mg of intravenous furosemide is administered to augment urine flow. It is important that the ureter is not divided until the last moment as this helps to prevent torsion of the kidney which could occur inadvertently. It may be difficult to rapidly untwist the kidney, leading to disastrous intraoperative ischemia.

If a Pfannenstiel incision is to be used for extraction, this is prepared now. A skin incision approximately equal to the diameter of the waist of the kidney is made through the fat to the muscle fascia, which is incised horizontally through the midline to the left side. The midline attachments to this anterior layer are divided with electrocautery as far inferiorly as the pubic symphysis and superiorly a similar distance. A 15 mm diameter trocar is inserted in the midline under direct vision from within. The only layer ensuring maintenance of the pneumoperitoneum will thus be the peritoneum posterior to the lower rectus muscles. The extraction bag device (EndoCatch II®, US Surgical, Norwalk, CT) is advanced through the trocar under vision, and fixed to the drapes so that it will not interfere with subsequent maneuvers (Fig. 4.12).

An 0-0 Vicryl® ligature is inserted through the 12 mm trocar (#2), looped around the ureter, tied extracorporeally and then snugged down around the ureter as far distally as possible, preferably at the pelvic brim. The scissors are introduced through the same trocar to divide the ureter and the ligating thread. A jet of urine indicates the kidney is functioning well. The kidney should be uniformly pink and firm. The loop around it is retrieved and discarded.

The donor is now heparinized.

The next steps depend on what technique is used for vascular ligation and are timed on the clock.

If a clip applier is to be used:

- The artery is doubly clipped at its origin on the aorta (Fig. 4.13).
- A delay of approximately 60 seconds will allow the kidney to completely empty through the vein which will be seen to collapse.
- The suture ligature around the vein is snugged tight. With a suction cannula positioned in the lateral 5 mm cannula, a straight scissors is used to divide the vein as the assistant sucks away blood that flows into the field. This will rapidly stop as the kidney empties.

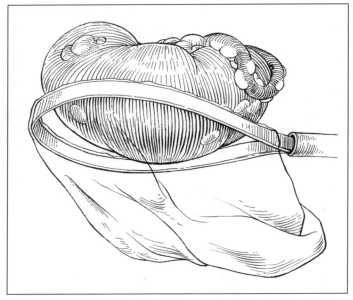

Fig. 4.12. Before division of the renal vessels the extraction incision is prepared, a 15 mm trocar inserted and the extraction device inserted ready for deployment.

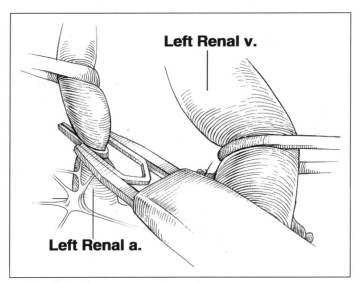

Left Renal v.

Left Renal a.

Fig. 4.13. Unless the artery is very short, clips (as here) or staples should not be applied flushed with the aorta. A short length of the stump proximal to the clips should be available for subsequent ligature.

- The same scissors then divides the artery. Vascular loops are set aside.
- The kidney is scooped into the fully deployed retrieval bag. A finger plunged through the Pfannenstiel incision on each side of the trocar stretches the posterior layer and the kidney is rapidly withdrawn through this incision and chilled. The warm ischemia time is recorded.

If a stapler is used the following sequence is necessary after test positioning of the stapler over vein and artery:

- Gentle lateral traction on the kidney straightens the pedicle. Upward or downward traction with the vein loop allows the stapler to be positioned over the origin of the renal artery. The tip is cleared of extraneous tissue and the stapler closed and fired. While the cartridge is exchanged a delay of about a minute will enable the kidney to empty.
- The vein is divided as far medially as is possible in a similar fashion.
- The kidney is extracted as above.
- Protamine is given to reverse the heparin (1 mg per 100 units of heparin).

4

RIGHT NEPHRECTOMY

The donor is positioned in left lateral decubitus. Lines are marked between the 12th rib and umbilicus, and in the infraumbilical site. Port placement is significantly different from the left side.

A 12 mm Hasson cannula will be sited in the right lower quadrant over the lateral border of the rectus and pneumoperitoneum induced. From this entry point later in the procedure the long axis of a laparoscopic Satinsky clamp will be roughly parallel to the cava allowing an optimal side bite at the take-off of the right renal vein. The 5 mm laparoscope and flank ports are inserted according to the landmarks for the left.

An epigastric (true paraxiphoid) 5 mm port is sited for a toothed grasper, used to retract the liver cephalad.

A 10 mm port is sited between the laparoscope and epigastric ports for a right-handed surgeon using the LCS. Alternatively a 5 mm cannula is possible only if scissors dissection is used.

Kidney extraction through a transverse, rectus-sparing incision sited just below the umbilicus reduces the chance of displacing the caval clamp during extraction. A 6 mm transverse infraumbilical incision is made in the skin and deepened to include the anterior and posterior rectus sheath, but preserving the longitudinal fibers of the rectus muscle by lateral retraction. A 15 mm port is inserted in the midline and with suitable reducing valves can be used for dissection, traction, suction etc. during mobilization.

If the 12 mm or 15 mm ports do not provide comfortable access to the renal hilum and upper pole, then a further 5 mm port can be sited at the lateral border of the rectus caudad to the laparoscope.

The exact configuration of these cannulae can be varied according to the surgeon's dominant hand, provided that adequate distance is maintained between working ports and the optical axis.

TECHNIQUE

First, the liver is elevated from Morrison's pouch. This may require division of the right triangular ligament of the liver if there is not enough mobility of the right lobe. A toothed grasper inserted through the epigastric port, under the liver and clamped high on the lateral abdominal wall will keep the liver elevated until the end of the procedure.

The hepatic flexure of the colon is reflected medially to expose the second portion of the duodenum. Inferiorly, enough mobilization of the colon to allow access to the ureter as it crosses the iliac vessels is necessary. The duodenum is partly Kocherized (reflected medially) by incision of the posterior peritoneum to expose the vena cava, right renal vein, anterior infrahepatic cava and gonadal vein origin. Dissection as high as the adrenal vein is not necessary. Gerota's fascia over the origin of the renal vein is incised longitudinally, staying medial to the gonadal vein, and the ureter identified and looped. The triangle of fat below the lower pole, lateral to the ureter and anterior to the psoas muscle is preserved to ensure good ureteral blood supply. Dissection proceeds superiorly along the upper border of the renal vein and upper pole, separating the fat containing the right adrenal from that surrounding the kidney. This is carried laterally over the top of the upper pole. The vein is freed circumferentially, looped and gently retracted as the pulsations of the renal artery are sought. The artery is dissected as on the left side, and a white vessel loop slung around it. Once the kidney has been fully mobilized from within the perirenal fat, it can be flopped medially and the arterial dissection completed from the posterior aspect. It is possible to dissect the artery well posterior to the cava, but as the limiting factor will be vein length, this dissection need not be prolonged.

As a safety measure a 3-0 Gore-Tex® pursestring is made around the origin of the renal vein (Fig. 4.14a), picking up three or four superficial caval bites on each side before being tied externally with a Roeder knot slipped down but not occluding the vein. This gives some measure of control to the side of the cava should the Satinsky clamp slip.

As it is an easy matter to ligate the artery with thread, we prefer to site an 0-0 silk tie around the artery, make a Roeder knot and slip it down behind the cava in preparation for final extraction.

Test positioning of the caval clamp ensures that all of the origin of the vein will be included with a little of the flare onto the cava and that the tips of the Satinsky are clear. The extraction wound is prepared by ensuring that the fascial incision will accommodate the kidney, but the peritoneum is left intact to be manually stretched on withdrawal of the EndoCatch. Meticulous hemostasis is now in order before heparinization. Once word is received that the recipient is ready, the ureter is ligated and divided and the donor heparinized.

An assistant retracts the kidney laterally as the arterial ligature is positioned and tightened. The Satinsky clamp is applied to the origin of the vein and clamped slowly tight, if necessary including the pursestring in the jaws (Fig. 4.14b). A straight scissors is used to cut first the vein and artery. It is critical to leave at least 2 mm of vein protruding above the jaws of the clamp for closure. Displacement of clips on the artery by Satinsky clamp, or inclusion in its jaws, could result in catastrophic

4

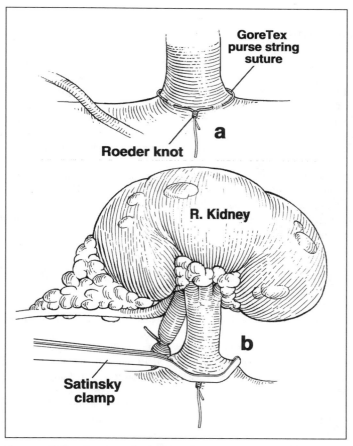

Fig. 4.14. a. Right nephrectomy: After dissection of the confluence of the right renal vein and the vena cava, a pursestring of 3-0 Gore-Tex® is fashioned at the point where the vein begins to flare onto the cava. This is tied with a Roeder knot which is slipped down close to the cava. b. The Satinsky clamp is inserted through a right lower quadrant 12 mm port. After ligature of the renal artery, the clamp is applied to the renal vein caval junction just inside the pursestring.

retrocaval hemorrhage that would be difficult to control, hence the use of thread ligation and deferral of division of the artery until the clamp has been applied. The kidney is scooped into the retrieval bag, extracted, chilled and the wound packed tightly enough with a sponge to allow maintenance of pneumoperitoneum. Protamine (1mg per 100 units of heparin) is given.

The vein is closed with a conventional running 3-0 or 4-0 Gore-Tex® suture tied intracorporeally, and the Satinsky clamp is slowly removed. The previously inserted pursestring can be gently tightened if close to the suture line, or merely cut short and left in place if tightening it threatens to tear the caval wall.

TERMINATION

The Pfannenstiel or infraumbilical incision is closed in layers with an absorbable suture material, the donor reinsufflated and laparoscope reinserted. There should be no bleeding whatsoever. Double clipping of an appropriately sized renal artery is safe, but should any concern about the pedicle remain, a pretied loop of 0-0 silk is used to secure it proximal to the clips or staples. Any clot is removed and both vein and artery loops are removed and discarded. On the left the spleen is replaced in the left upper quadrant, but no attempt is made to fix it or the colon to the parietes. On the right the liver retraction clamp is removed and the muscle and liver inspected for bleeding or damage. The posterior aspect of the extraction wound is inspected to ensure that nothing has been picked up during closure. All 10 or 12 mm punctures should be closed at the fascial level.

Subcuticular skin closure of the longer incisions will help approximate the dermis to facilitate application of sterile adhesive strips.

AFTER CARE

As there will be a substantial diuresis, the urinary catheter remains indwelling until the first postoperative day when it is removed at 6 AM. A single dose of parenteral opiate analgesic is given in the recovery room, and the patient commenced on clear fluids the evening of surgery. Oral analgesics can then be administered. Remarkably few analgesics are required. Patient controlled analgesic infusions are not indicated and if used will guarantee slower return of physical activity and intestinal function.

Calf compression boots are removed and the donor encouraged to ambulate as soon as possible. Prophylactic heparin injections continue until discharge. A regular diet is ordered the following day and the donor encouraged to take free oral fluids. An intravenous line may or may not be necessary at this stage. Abdominal distention by the second postoperative day is common, but the donor should soon pass gas. There is no need to wait for a bowel movement before allowing the donor home. Fever, dizziness or increasing abdominal pain should be reported to the surgical team.

With this regimen most donors will be fit for discharge on the second postoperative morning. The donor may shower from day one, but the sterile adhesive strips should be left in place to be removed in the office at the end of the second postoperative week. Multiple in-hospital laboratory tests are not indicated if the donor is stable with good urine output. A single measurement of hematocrit, serum electrolytes and urea/creatinine are performed on the second postoperative day. These are re-evaluated at the first office visit.

COMPLICATION PREVENTION

With meticulous technique, an experienced operator with good laparoscopic skills should be able to perform this operation without injury to other organs.

ON THE LEFT SIDE
- The spleen is at risk from capsular tear if roughly retracted or from parenchymal puncture by a carelessly inserted grasper in the most cephalad port.
- Handle the colonic adnexae and not the colon wall to avoid a full-thickness injury.
- Be aware of the proximity of the gastric fundus during division of the uppermost part of the splenophrenic ligament to prevent a gastric injury.

ON THE RIGHT SIDE
- Open laparoscopy will avoid injury to the gallbladder with the Veress needle.
- Handle the peritoneum of the duodenum or take a large bite with a gentle atraumatic grasper of the duodenal wall during Kocherization.

GENERAL
- Keep the tip of the laparoscope meticulously clean. The commonest mistake is to proceed with poor image quality causing frustration and premature fatigue to the operative team.
- Avoid stripping the ureter bare and preserve the fatty triangle below the lower pole of the kidney to prevent ischemic necrosis of the ureter.
- Divide the ureter just before extraction to prevent torsion of the kidney on its vascular pedicle.
- When dissecting blood vessels the same principles as in vascular surgery apply: stay close to the vessel with sharp dissection, counter-retract using adventitia or Vessiloops®. Moistening the area with a papaverine soaked gauze roll can reduce spasm, although good technique is more important.
- Apply clips securely enough to arteries that they will not slip off easily, but not so firmly that they cut through the vessel wall.

- Position clip appliers or staplers to leave 1 mm or more of accessible vessel stump. Tying a subsequent ligature or stopping bleeding at the site of the clips or staples is much easier.
- Stop all bleeding before proceeding to the next step of the operation. Pooling blood absorbs light and reduces image definition, making accurate surgery impossible. Not only is this good surgical technique, but reduces overall blood loss, particularly during anticoagulation.
- A gauze roll, tied at each end, with one end tagged with a long thread that trails out through a cannula is indispensable for pressure hemostasis, blunt dissection and generally mopping up. The thread facilitates retrieval of the roll.
- Vessel loop retraction of vessels is elegant and gentle but should be for short periods only to prevent ischemic insult to the kidney. Blood flow measurements show that retraction sufficient to create an angle in the renal vein causes almost instantaneous cessation of flow in the renal artery.

CONVERSION TO LAPAROTOMY

Because bleeding from a major vascular structure is the greatest risk, it is wise to perform this procedure with a full laparotomy setup open in the operating room. The line drawn from rib to umbilicus can be incised quickly without desufflation should bleeding from a major vessel be uncontrollable with gauze roll pressure. Major bleeding will usually occur after a good deal of the exposure (colon, spleen reflection) has already been done, and access will not be restricted. Little will be achieved by prolonged pressure on a feeding vessel to what is now an ischemic kidney. The chief priority is protection of the donor, so that if the question arises as to whether to convert, that is probably the best course.

There are few anatomic reasons for conversion. Perhaps the most obvious would be an extremely short right renal vein that did not permit a long enough cuff beyond the clamp for suture closure. An unidentified circumaortic left renal vein may prove anatomically too challenging for the laparoscopic technique. Double ureters, arteries and veins, if identified on preoperative imaging can point the surgeon to remove the other side, but pose no special problem when recognized at operation, provided that smaller vessels are suitable for anastomosis. At present taking a cuff of aortic wall is beyond the scope of laparoscopic technique.

The End Stage Renal Disease Patient and Evaluation for Transplant

5

Maureen Ulrich, Laura Coorpender, Karen Paolini, Susan P. Graham

BACKGROUND

Transplantation is one part of the overall continuum of health care for patients with end stage organ failure. The process begins with a referral to the transplant center for evaluation and continues for the duration of the organ function or the life of the patient (see chapter 1, Fig. 1.1). The transplant procedure goes beyond the 3-4 h of surgery time, representing a complicated process which can yield optimal results only with support and commitment from the patient and their support systems and the multidisciplinary team that manages the patient's care. This team is comprised of transplant surgeons, transplant nephrologists, referring physicians, transplant coordinators, the hospital transplant unit, consultants, dietitians, the operating room, histocompatibility laboratory, hospital administration, the organ procurement organization and many other ancillary facilities. Once the patient is discharged, the outpatient clinic and home care services are incorporated, depending on individual needs and plan of care. In order to achieve long-term success of a transplant, the multidisciplinary team must work together, with each team member clear in regards to their specific role. Throughout the transplant process, this team should be responsible for overseeing the coordination of all interventions for the transplant patient to ensure continuity of care.

Kidney and Pancreas Transplantation, 2nd Edition, edited by Ingemar J.A. Davidson © 1999 Landes Bioscience

PROTOCOLS IMPROVE OUTCOME

The success of transplantation is influenced by the establishment and utilization of protocols based on scientific research and proven methods of providing safe and effective treatment. When protocols are followed closely, there is a greater chance for a successful outcome. Because of the complexity of the transplant process, it is necessary to have protocols, check lists, standardized procedures and clinical practice guidelines. All of these will ensure that patient safety is maintained and that the process is completed in an efficient and cost effective manner. Strict implementation of protocols and procedures will greatly reduce errors and facilitate early identification of complications. Communication between all persons or departments involved in the transplant process is very important. Additionally, the transplant program needs to have a strong working relationship with the hospital administration. Without this support it will be difficult for the transplant program to effectively implement the protocols and to provide high quality care. It is the hospital administration that determines how much space will be dedicated to transplant, the number of nursing staff required and support for the transplant program from departments within the institution. These and other administrative decisions will impact outcomes, particularly if there is not adequate support for the transplant program. Frequent and open communication with the hospital administration as well as mutually agreeable goals in regard to the transplant program are necessary to make the program stronger and more competitive in the transplant community.

The similarity to the airline industry is striking. Often, a series of errors or mishaps (miscommunications) lead up to a single fatal decision or event, and, frequently, the responsibility rests on a single person. This individual is usually the pilot in the case of an airline, and a surgeon or anesthesiologist in the operating room setting. Each team member's duties and responsibilities must be defined. Because so many departments and institutions are involved throughout the entire transplant process, key members must be identified and given the responsibility to coordinate, streamline and implement protocols consistently. A successful transplant program is dependent on each member of the team performing their responsibilities without fail, as well as adhering to a positive system of checks and balances which keeps the entire team on track.

Protocols will markedly vary between transplant programs because of local resources, research protocols, institutional commitments and personal experience and preferences. Protocols must be continuously updated as medical knowledge and technology evolve. Successful development and implementation of protocols is very program dependent; however, responsibility will generally rest with a few key individuals. For example, what may be a routine process in one transplant center requires regular maintenance in another. Each time a protocol is updated or changed the date should be noted, the changes distributed and communicated to all members of the transplant team. Old protocols should be discarded in order to avoid confusion.

A clinical practice pathway can provide a method to track outcomes relative to the protocols and procedures. This guideline is a recommendation for patient management, which is built from the protocols. Individual patient's care may vary from the guideline, but documentation should be provided as to why the patient's care was different from the clinical practice pathway. As a result, the transplant program will have data about the expected outcomes and variances to the expected outcomes. This will enable the program to determine if changes in the protocols need to occur so that inappropriate care is reduced and patient outcomes improve.

WHEN TO REFER FOR TRANSPLANT

5

The patient facing end stage renal disease (ESRD) has three treatment options. He or she can choose one of two dialysis options: hemodialysis or chronic peritoneal dialysis. The third option is to consider a transplant, where there are two alternatives: a living donor transplant, usually from a close relative, or a cadaver renal transplant.

Pre-emptive transplantation (before dialysis is required) is a preferred option, although seldom realized because of a shortage of organ donors. Another limiting factor in pre-emptive transplantation is the delay in referring patients to a nephrologist for ESRD care, or nephrologist delay in referring to the transplant team. Ideally, patients should be referred to the transplant program upon diagnosis of impending renal failure, before the need for dialysis treatment is a reality. The referral of the patient can occur by a variety of sources, i.e., by self referral, dialysis unit nurse, primary care physician, family member or insurance companies. The timing of the transplant evaluation is largely dependent on the nephrologists' referral pattern. As provided by the Health Care Finance Administration, all patients on dialysis must be reviewed annually for the medical suitability for transplantation. Early referral is the key to successful and cost effective ESRD management. For example, in the case of a recipient with a living donor, the transplant procedure may be timed to avoid dialysis therapy. Regardless of the availability of a living donor, the success of transplantation is greatly enhanced when the recipients are in good condition at the time of the transplant procedure (chapter 6, Table 6.2). Timely transplant evaluation increases the likelihood of an early and successful transplantation. It also allows the patient to see transplant as a positive option for ESRD rather than an alternative to undesirable or unsuccessful dialysis. The patients' insurance status can influence the transplant process. The future trend is for hospitals and insurance companies to enter into contractual agreements for transplant services, making verification early in the process critical.

PATIENTS APPROACHING END STAGE RENAL DISEASE (ESRD)

Vascular access for hemodialysis or Tenckhoff catheters for chronic peritoneal dialysis should be placed well prior to the need for dialysis to allow maturation. This timing depends on the individual's predicted decline of renal function. As a rule of thumb, diabetic patients should have their access placement performed when their creatinine level is 4-5 mg% (GFR = 15-20 ml/min). Other stable patients with no significant contributing medical problems should have access placement when the creatinine approaches 7-8 mg% (GFR = 10-15 ml/min.). Patient and family education is critical in the success and timely management of ESRD patients in preparation for transplantation. Thoughtful planning in regard to dialysis access placement will prevent the need for central vein dual lumen catheters or uncomfortable femoral vein cannulations. Dual lumen catheter placement is costly, in terms of both finances and central vein morbidity. The preservation of vascular integrity is vital in the ESRD population. As a general rule, if a dual lumen catheter is indicated, the internal or external jugular vein should be the site of choice, because of a lower incidence of central vein stenosis. These should be placed preferably under the guidance of portable sonography. Percutaneous subclavian catheters should be limited to emergency situations, due to the high incidence of serious vascular complications, in addition to pneumothorax and hemothorax.

Transplantation is the preferred choice in medically suited patients. It markedly improves the quality of life and is also cost effective for society, as compared to patients who remain on dialysis treatment. The time line of the ESRD patient was described in more detail in chapter 1, Figure 1.1.

EVALUATION OF ESRD PATIENTS FOR TRANSPLANTATION

All potential candidates for transplant must undergo pretransplant evaluation. The overall purpose is to determine that indications for transplant exist and that the patient meets the medical criteria indicating that the transplant procedure can be performed safely with a high likelihood of long-term success. Should medical problems or contraindications be identified, these should be corrected prior to transplantation.

Standard established criteria for kidney transplantation include:
- presence of or impending ESRD
- acceptable medical/surgical risk
- absence of end stage or severe extrarenal disease
- social and/or psychological stability
- absence of infection or infectious disease process
- absence of a malignancy or resolution > 2 years

For a simultaneous pancreas/kidney transplant:
- presence of type I diabetes (as evidenced by lack of C-peptide production)
- absence of extensive disease related to diabetic complications

TRANSPLANT TEAM MEMBER FUNCTIONS

The purpose of protocols in the pretransplant evaluation is to establish comprehensive criteria that will provide an effective, thorough evaluation tool for a wide variety of patients with equally varying risks and needs. The most effective transplant process will begin with a detailed review of each potential candidate's history, including review of medical records, followed by an individual evaluation process based on the patient's history. This process is most efficiently and cost effectively put into place by the transplant coordinator. The final decision regarding transplant status is ultimately the decision of the transplant surgeon, this decision must take into consideration the evaluation done by the transplant coordinator, the referring and transplant nephrologists, social workers, and consulting physicians. Equally important is an objective assessment of the medical and psychological data revealed through diagnostic transplant evaluation (see Table 5.1). A transplant physician's most vital tool in the evaluation process is a comprehensive history, a thorough physical examination and the ability to incorporate the information obtained through the transplant evaluation and to make a recommendation for each individual based on a risk vs. benefits model. In addition to a patient's risk for the surgical procedure itself, the team must recognize the many factors which may make the patient an unacceptable risk for long-term immunosuppression. Abnormal findings in the evaluation will be considered based on potential impact on transplant outcome. It is critical to develop the skill and judgment based on personal experience as well as the experience of colleagues with regard to weighing the risk vs. benefits, as well as with regard to the recommendations for further diagnostic studies or consults. Although many of the criteria regarding candidacy for transplant will vary between transplant centers, there are several absolute contraindications to kidney and/or kidney/pancreas transplantation that are generally accepted nationally (Table 5.2).

Additional diagnostic studies may be completed at a facility other than the transplant center as part of a routine screening. For a variety of reasons, we have chosen to recommend these additional studies for specific categories of patients based on individual history or clinical presentation. These decisions are made in an effort to create a transplant evaluation which will provide the pertinent information, remain cost effective and avoid subjecting patients to unnecessary risks or burden associated with invasive diagnostic testing. Examples of these additional studies and suggested indications are listed in Table 5.3. The decision to

Table 5.1. Common laboratory and diagnostic evaluations for all potential transplant candidates

A. Laboratory
    Amylase and Lipase
    Chemistry Panel
    Liver Enzymes
    Cholesterol Profile
    PT/PTT
    ABO Typing
    HLA Typing
    Pretransplant Crossmatch
    TB Skin Test
    Urinalysis
    Urine Culture
    Serology Panel
     HTLV-1
     Hepatitis Panel–A,B,C
     HIV
     CMV IgG/IgM
     EBV IgG/IgM
     HSV I IgG/IgM
     HSV II IgG/IgM
     Mumps Antibody
     Rubella Antibody
     Rubeola Antibody
     Varicella Zoster
     RPR
B. Diagnostic Studies
    Chest X-Ray
    EKG
    Cardiac Echo
    Abdominal Ultrasound
C. Consultations
    Transplant Surgeon
    Transplant Nephrologist
    Transplant Coordinator
    Transplant Social Worker
    Dental Evaluation
    Gynecological Evaluation
     Including PAP Smear, Mammogram

pursue more extensive diagnostic testing is a team decision based on identification of risk factors which may adversely affect transplant outcome. Further exploration may therefore allow for more precise decision making.

MAINTAINING PATIENTS WHILE ON THE WAITING LIST

Once a patient is accepted for transplant and placed on the UNOS waiting list, it is imperative that the transplant team remains informed regarding the patient's

Table 5.2. Absolute contraindications to renal or pancreas transplantation

Presence of active infection or sepsis
Open wound
HIV positive
Cancer without a 2-5 year remission period (Remission time required based on
the type of treatment and the risk for recurrence)
Hematologic cancers or lymphoma
Severe or irreversible extrarenal disease (i.e. heart, lung, liver)
Unacceptable vascular integrity
Unresolved social or psychiatric/psychological factors (i.e. drug addiction)
Evidence of **active** systemic disease (i.e. lupus, Wegener's, sickle cell)
Morbid obesity (Weight > 30% above ideal body weight)
Positive T cell crossmatch
Active hepatitis
Peritonitis or graft infection without 2-4 weeks resolution

5

changing status, as it is impossible to predict how long a patient will wait for a
cadaver kidney. Waiting time varies from center to center and is based on a patient's
ABO group. An additional factor affecting waiting time is the presence of preformed reactive antibodies (PRA). The better means for estimating time on the
waiting list is the median (i.e., the time in which half of the patients would be
transplanted). An example of median waiting time at SUNY Buffalo is given in
Figure 5.1. It should be noted that patients who currently wait a long time typically have had previous transplants or blood transfusions, resulting in high PRA
levels often associated with incompatible crossmatches (see chapter 7). Whereas a
patient having blood type B may typically wait longer because of the rare donor
situation, a patient in the O blood group may wait longer due to a larger number
of candidates awaiting transplant in that blood group. During the waiting process, regular and continuous communication with the patient, dialysis unit and
nephrologist is recommended. Ideally, the patient should visit the transplant clinic
every 6-12 months while awaiting a kidney transplant and every 3-4 months when
diabetes is the etiology of ESRD or if awaiting pancreas transplant. At these visits,
the transplant team should look for new events that would potentially temporarily rule the patient out, i.e., EKG changes, ischemic foot ulcers, angina, and also
continue the patient/family educational process. Additionally, factors which may
render a patient in more urgent need of transplant should be considered and reviewed by the transplant committee.

CARDIAC EVALUATION FOR KIDNEY/PANCREAS TRANSPLANT

Patients referred for kidney and kidney/pancreas transplantation are, by definition, a high risk group of patients. All patients for kidney/pancreas transplant
have had long standing diabetes, which has led to end organ damage. Additionally, many patients for kidney transplant also have diabetes mellitus as an etiologic

Table 5.3. Additional extensive evaluation based on specific individual patient indications

Indication	Age> 50	Clinical History	Abnormal Prior Test	Type of Abnormal Test	SPK Consideration	Smoking History
Test						
Mammogram	•	•	•	Clinical exam		
Cardiac Stress Test		•	•	Echo, EKG	•	
Coronary Angiogram		•	•	Previous CAD, Abnormal PET		•
VCUG	•	•			•	
PET			•	Echo, EKG, Stress Test		
Drug Rehabilitation		•	•	Random Drug Screening		
Urology Consult		•	•	Abnormal/Unsuccessful VCUG		
Psychiatry Consult		•				
GI Consult		•	•	US/Abdominal CT/Clinical exam		
Liver Biopsy		•	•	Abnormal CT/Hx of Hepatitis		
Vascular Dopplers	• >60	•	•	Clinical exam, Symptomatology	•	•
Barium Enema		•	•	+ Stool heme		
Colonoscopy		•	•	GI History, + stool heme		
PFT	• >55	•	•	CXR/CT Scan/Clinical Exam		•
Pulmonary Consult	• >55	•		•		•
INH Prophylaxis		• TB	•	+ TB skin test		
C–Peptide					•	
PSA	•	•	•	Clinical Exam		
Abdominal CT		•	•	Hepatitis +, Abnormal US		
Dietary Counseling		•	•	Weight > 30% above IBW		
PCR		•	•	+ HIV, HTLV1, Hepatitis C		
Cholecystectomy		•	•	Symptomatic/Abnormal US		

Clinical exam–history and physical, physical exam, laboratory evaluation; VCUG–voiding cystourethrogram; PFT–pulmonary function test; PET (positron emission tomography) may be requested in place of coronary angiogram. PSA–prostatic specific antigen; PCR–polymerase chain reaction; CXR–chest x-ray; IBW–ideal body weight; US–ultrasound
*It is the authors' bias that presence at Doppler studies improves the overall understanding of the vascular integrity relative to the surgical procedure.

factor. The process of kidney failure leads to hypertension which further accentuates the cardiovascular risk for surgery. Many patients may have poor ability to exercise due to bony or neuropathic processes, thus limiting the information available from a history about exercise induced symptomatology. Silent ischemia is an additional concern which may be present in any individual, but has a higher prevalence in patients with diabetes. Cardiac event remains a leading cause of complications and death in diabetic transplant recipients.

HISTORY AND PHYSICAL

The first step in risk assessment involves a careful history and physical examination. The history should include specific attention to the following known coronary risk factors, plus any prior history of cardiac, cerebrovascular or peripheral vascular disease; diabetes mellitus, including duration, control and organ injury; hypertension; cigarette smoking, current and past; lipid status, including treatment; and family history of coronary artery disease.

ASSESSMENT OF PHYSICAL ACTIVITIES AND EXERCISE TOLERANCE

The history should try to discern whether the patient has had heart failure independent of volume status from the renal insufficiency. It is often difficult to differentiate between volume excess and symptoms related to ischemia. However, a history of dyspnea on exertion the day after dialysis or dyspnea on exertion in a patient maintained on CAPD at dry weight should be highly suspicious for angina (dyspnea on exertion being an anginal equivalent). Patients may specifically deny chest pain, but may acknowledge symptoms such as chest pressure, heaviness, tightness or an aching in their arms, neck, jaw, epigastrium or chest area. These symptoms would be classified as anginal equivalents.

STEPS IN CARDIAC EVALUATION

- Baseline EKGs should be obtained. Comparison with old EKGs is very helpful when the EKG is abnormal.
- The next step in cardiovascular evaluation is to assess left ventricular function. Left ventricular function (ejection fraction) will be affected by: the underlying disease, the presence of uremia and poor metabolic control concomitant illnesses, especially viral infections and occult coronary artery disease.
  Ejection fraction can be measured either by a good quality echocardiography or by radionuclide imaging (MUGA scan).
- Provocative Testing: Patients undergoing major surgery need provocative testing to evaluate for the presence of occult coronary artery disease. There are a number of different ways of evaluating the coronary circulation and the type of testing will depend on the availability and quality of testing modalities at a particular institution. Provocative testing should provide information on the presence of obstructive coronary artery disease that will influence coronary flow and thus induce ischemia during the stress of the perioperative period. Unfortunately,

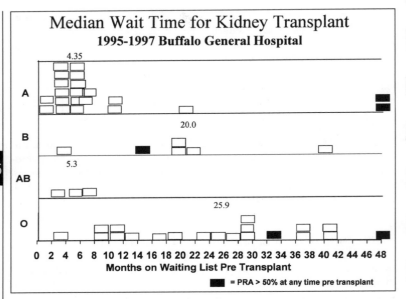

Fig. 5.1. Time on waiting list for cadaver renal transplant (CRT) for each ABO blood type. The median time for each group is depicted. Typically, patients with high panel reactive antibodies (PRA) wait longer.

one cannot test for the potential of cholesterol plaque rupture. This is the most common etiology of acute myocardial infarction in an asymptomatic patient and will be addressed below.

Provocative testing includes: stress test, nuclear stress test, pharmacologic nuclear stress test, stress echo with exercise of pharmacologic agents and PET (positron emission tomography). Valuable information is obtained about the patient's physiologic reserve during exercise stress testing. However, the sensitivity and specificity of this modality is relatively low. Pharmacologic stress testing (adenosine, dipyridamole, dobutamine) is hampered by the fact that renal failure patients have high levels of adenosine. Thus, the incremental increase in blood flow achieved by these pharmacologic agents is less than in the nonrenal failure population where the tests have been studied and validated. Traditional and nuclear stress testing is insensitive for the evidence of small vessel disease which may be present, especially in diabetic patients.

PET scanning is not available at every institution and may be a valuable source of information. PET scanning can demonstrate the presence of obstructive coronary lesions, quantitate the distribution of myocardium affected and also give information about coronary flow reserve and potential effects of small vessel disease. PET scanning may be a useful modality to follow in a serial fashion, along with aggressive efforts of lipid lowering to modulate the negative effect of advanced metabolic disease on coronary blood flow. Whether reduction of lipids will stabilize and eliminate the possibility of plaque rupture has not yet been proven, but is certainly a possibility and a hypothesis to be tested in the high risk population.

• Coronary angiography will precisely define coronary anatomy, but provides no information about the physiologic significance of stenosis. Angiography is not practical in a patient who has advanced renal insufficiency but is not yet on dialysis, since the test itself may precipitate complete renal failure. This may be an unacceptable option in patients who have not yet been accepted for transplant and are trying to avoid dialysis.

5

PLAQUE RUPTURE

Cholesterol plaque lines all arteries. The degree of friability and vulnerability of plaque to rupture is a function of many variables including genetic inheritance, metabolic milieu and lipid status. Aggressive efforts at lipid lowering have been shown to reduce cardiovascular events such as unstable angina and acute myocardial infarction in patients with established coronary artery disease. Before and after transplantation, every effort should be made to evaluate the patient's lipid status and consider methods to treat patients with an abnormal lipid profile. One should also be aware that plaque rupture may occur despite one's efforts, before, during or after the surgery and the health care team must be on alert for this possibility. Aggressive treatment including direct intervention with angioplasty should be attempted. Typically, plaque rupture will involve single vessel disease which may be amenable to revascularization but not require immediate coronary artery bypass grafting. Follow-up aggressive treatment of lipids is also strongly recommended, although the treatment becomes more complicated due to drug interaction with immunosuppressive agents. In the long run, this is an area of great potential to reduce the high cardiovascular mortality and morbidity following successful kidney and kidney/pancreas transplantation.

LOW EJECTION FRACTION

Patients with renal disease may have a low ejection fraction for several reasons: uremia, small vessel disease, intermittent illness or atherosclerotic coronary artery disease.

Patients who are under dialyzed will often have a decrease in ejection fraction with global hypokinesis which may be reversed with aggressive dialysis. There are numerous examples of patients whose ejection fractions have improved from 20% to near normal with alteration of their dialysis regime. Thus, the patient who has a low ejection fraction should first be subjected to critical evaluation and modification of dialysis before excluding them from transplantation. Patients who are uremic but not yet on dialysis may warrant initiation of dialysis for reversal of this problem.

SMALL VESSEL DISEASE

Patients with diabetes and hypertension may have "small vessel disease." Other manifestations of small vessel disease include, but are not limited to diabetic neuropathy, gastropathy, ophthalmopathy and nephropathy. The presence of one organ injury does not completely correlate with cardiovascular disease, but is suggestive of that etiologic pattern. The conclusion that a patient has small vessel coronary disease may occur because of: normal coronary arteries with low ejection fraction and otherwise good control of their renal failure, reduced coronary flow reserve on PET scanning, patchy abnormalities on nuclear imaging with angiographic normal large epicardial coronary arteries and heterogenous response to dobutamine stress echo.

The treatment of small vessel disease is to optimize metabolic control of diabetes and hypertension and focus on lipid treatment. These patients should NOT be automatically excluded from transplantation.

Concomitant illnesses such as a viral infection can often transiently reduce the ejection fraction. Before excluding the patient from transplantation due to low ejection fraction, the patient should have at least two measurements of ejection fraction at separate times and separate conditions to search for a fluctuating ejection fraction.

The transplant evaluation is outlined with an understanding of the posttransplant complications (Table 5.4). Individuals who are at most risk for

Table 5.4. Potential complications after transplantation

Vascular	Thrombosis
	Stenosis
	Bleeding
Wound	Infection
	Lymphocele
	Hematoma
	Poor healing
Ureteral/bladder	Leak
	Obstruction
	Fistula formation
Infection	Bacterial
	Viral
Malignancies	Skin
	Uterine/cervical
	PTLD
	Kaposi's sarcoma
Rejection	Acute vascular
	Chronic
Cardiac	MI
	Hypercholesterolemia
Bone	Fractures
	Osteoporosis

Most severe complications after transplantation are related to immunosuppression

serious complications are identified and properly advised and counseled. It is important to practice transplantation with the understanding that a certain level of risk is inevitable. The goal is to perform an effective transplant evaluation that will provide a realistic identification of risks while remaining cost conscious and fair to patients.

LIVING DONOR EVALUATION

Once a patient is determined to be an acceptable recipient, the coordinator begins the process of identifying an acceptable living donor. The process of evaluating potential living donors should not begin until the recipient has been accepted as a candidate. An area of controversy is the importance of HLA recipient-donor matching for outcome. It is generally accepted that in the case of six antigen matched kidneys there is a decreased incidence of rejection and about 10% better chance for long-term graft survival. Basic HLA evaluation includes identifying specific Class I genes (HLA–A&B) and Class II genes (HLA-DR) (see chapter 7) and the matching with potential cadaver donors is based on the UNOS guidelines. UNOS guidelines implement a national system by which donors are matched to recipients based on several recipient driven factors, i.e., ABO group, HLA type, % PRA, time waiting). The importance of less antigen matching remains controversial. The immunosuppression protocols outlined in chapter 6, Table 6.10 provide early, consistent and effective use of immunosuppression which results in an incidence of early episodes of rejection of about 10% (chapter 6, Fig. 6.3, 6.7b), thus minimizing the importance of HLA matching. Additionally, in cases where recipient underlying renal disease has the potential to re-occur, i.e., IgA nephropathy, hemolytic uremic syndrome (HUS) or most commonly, focal segmental glomerulonephritis (FSGN), transplantation with a living donor or well matched cadaver donor may promote re-occurrence, making a less well matched organ desirable. Early use of high CsA blood levels of 300-400 ng/ml (TDx/ TDxFLx® Cyclosporine Monoclonal Whole Blood Assay, Abbott Laboratories., Abbott kit) may also prevent reoccurrence of, i.e., FSGN or IgA nephropathy.

The living donor evaluation includes a physician evaluation with a thorough history and physical, nephrology medical clearance, consultation with the transplant coordinator and the transplant social worker, chest X-ray, EKG and an abdominal ultrasound, full laboratory panel similar to the recipient work up, with the addition of a 24 hour urine creatinine clearance and protein, an intravenous pyelogram (IVP) and renal arteriogram. In order to remain cost conscious and to avoid subjecting a patient to unnecessary testing, the donor evaluation is completed in a series of steps beginning with confirmation of ABO compatibility, followed by a preliminary crossmatch, chemistry panel and a 24-hour urine sample for protein and calculated creatinine clearance. The living donor candidate is usually not financially responsible for the evaluation. This is usually covered by the recipients' insurance. Currently accepted practices do not allow for financial reimbursement or provisions for the living donor or cadaver donor family, although

5

discussion on these issues has increased on a national level. Most living donor programs have expanded their criteria for acceptable donors to include "emotionally" related donors. This remains a controversial issue, but is undoubtedly a creative way of compensating for the organ donor shortage.

SELECTED READING

1   Donation and Transplantation: Nursing Curriculum. Chabalewski F, ed. UNOS, Richmond, VA, and the Division of Transplantation, Bureau of Health Resources Development, Health Resources and Services Administration, US Department of Health & Human Services, Rockville, MD.

2   Colombe BW, Garovoy MR. The major histocompatibility system of humans. In: Jacobson H, Striker GE, and Klahr, S, eds. The Principles and Practice of Nephrology. St. Louis: Mosby, 1995 (second edition); 798-805.

3   First MR. Pretransplantation evaluation and preparation of donors and recipients. In: Jacobson H, Striker GE, and Klahr, S, eds. The Principles and Practice of Nephrology. St. Louis: Mosby, 1995 (second edition); 805-10.

4   Kasiske BL, Ramos EL, Gaston RS et al. The evaluation of renal transplant candidates: clinical practice guidelines. Patient Care and Education Committee of the American Society of Transplant Physicians. J Am Soc Neph 1995; 6(1): 1-34.

5   Leape L. Error in medicine. JAMA. 1994; 772(23): 1851-57.

6   Park JW, Merz M, Braun P. Regression of transplant coronary artery disease during chronic low density lipoprotein apheresis. J Heart Lung Trans 1997: 290-7.

7   Penn I. Tumors in the transplant patient. In: Jacobson H, Striker GE, and Klahr, S, eds. The Principles and Practice of Nephrology. St. Louis: Mosby, 1995 (second edition); 833-37.

8   Sagalowsky AI, Dawidson IJA. Surgical Complications of Renal Transplantation. In: Jacobson H, Striker GE, and Klahr, S, eds. The Principles and Practice of Nephrology. St. Louis: Mosby, 1995 (second edition); 822-26.

9   Woolf SH. Practice guidelines, a new reality in medicine. II–methods of developing guidelines. Arch Int Med. 1992; 152:946-52.

10   Woolf SH. Practice guidelines, a new reality in medicine. III–impact on patient care. Arch Int Med. 1993; 153:2646-55.

Transplant Protocols and Outcome

Ingemar J. A. Davidson, Carmencita Cava-Bartsch, Christopher Lu,
Carolyn Munschauer, Maureen Ulrich

6

PERIOPERATIVE TRANSPLANT COORDINATION

After the identification and notification of a potential cadaveric renal transplant candidate has been completed, the next step in the preoperative transplant process is the medical clearance and preparation of the patient for surgery. The primary motivation to complete this process quickly is to shorten the cold ischemia time, repeatedly implicated as the prime factor in early renal function, impacting graft and patient survival. Also, an increasing body of literature indicates that acute renal failure of the allograft (i.e. delayed function) is associated with increased incidence of rejection and poor outcome. Therefore, the coordination of the operative transplant procedure involves facilitating the patient's arrival in the operating room in the shortest time possible, with all parts of preoperative workup complete.

Due to the nature of transplantation and the desire to minimize ischemic time, many patients arrive at the hospital during night hours, when outpatient departments are closed and staff are not available. The availability of a transplant team member guarantees the completion of preoperative and departmental protocols, the movement of patients through the hospital system and the identification and solution of particular impediments to arrival in the operating room (Fig. 6.1).

While each step in the preparation for surgery may be the immediate responsibility of an individual department, the stepwise progression of evaluations and results reporting is best overseen by a transplant team member communicating with departments and staging the patient's progress through the system. In this way, this individual functions as a systems administrator, interacting with the operating room and preoperative units. Depending on program structure, this individual can be a transplant coordinator, a medical or surgical house officer, transplant fellow or someone familiar with the local protocols and hospital system.

A decision must be made early whether the recipient will need dialysis before the transplant surgery. Assessment begins at the first phone contact with the patient based on the time lapse from their last dialysis, weight gain and other factors.

Effective execution of protocols requires timely communication with the operating room, pharmacy, radiology, intensive care unit or postanesthesia unit, anesthesia department, laboratory services and the admitting unit, i.e., the emergency room, day surgery, ambulatory admission unit or renal floor (Fig. 6.1).

The preoperative process involves review of history and physical, laboratory results, EKG, chest x-ray and review of the transplant chart. The initial step before the patient can go to the operating room is verification of crossmatch results (Fig. 6.1). A negative crossmatch means that the donor kidney (organ) is compatible with the recipient and that so-called hyperacute antibody-mediated rejection will not occur. Also, early in the crossmatch testing a discussion with the tissue typing laboratory should determine whether the potential recipient needs more sophisticated testing such as flow cytometry or B cell crossmatch. This is especially pertinent when the recipient has a high preformed antibody level or antibodies to known HLA antigens that may have developed from previous transplants, blood transfusions or pregnancies in females (see Chapter 7).

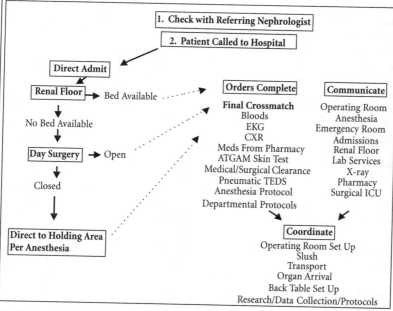

Fig. 6.1. Examples of preoperative hospital routes in cadaver renal transplant recipient work-up depending on time of day, transplant program structure and administrative support.

The operating room set-up consists of specially prepared transplant instrumentation pans, an intraoperative fluid and drug administration protocol adjusted for each patient and a self-contained unicell providing immediate access to instruments, sutures and supplies (Chapter 1, Fig. 1.2). The operating room must be set-up to accommodate first the back-table organ preparation, then the procedure itself, including the availability of sterile slush and additional supplies.

Once the transplant procedure begins, data relevant to the procedure, including anastomosis times, cross-clamp time, fluid volume and intraoperative medications administered are logged, and data for departmental research projects is collected.

Table 6.1 summarizes some general medical and local program system factors that influence transplant outcome. Table 6.2 highlights local factors to consider when building or improving a transplant program. Although these tables are self-explanatory, a few comments are warranted. Maintenance of the waitlist includes close communication with the referring nephrologist, who is usually also the patient's dialysis doctor. The better the general health of the patient, the better the outcome of transplantation. Yearly re-evaluations at the transplant center for uncomplicated recipients are recommended, with more frequent follow-up for diabetics and older patients. Part of the success of a transplant program is to educate the referring physicians and dialysis centers as to the needs of the transplant team to remain updated on the candidates' condition while waiting for transplant.

PREOPERATIVE ORDERS

Once a donor organ has been accepted and the recipient list generated by UNOS is reviewed, the best potential candidate is chosen and the admission process begins. First, the candidate is screened by phone to ensure that there are no current or new contraindications to the transplant procedure, such as active infections, especially toe or foot infections in diabetics, fever, cough, or cardiac events since

---

Table 6.1. Some important medical factors influencing transplant outcome

1. Optimal donor management (fluid, drugs)
2. Preprocurement nodes for HLA and crossmatching. (shortens CIT*)
3. Optimization of multi-organ donor management and procurement surgical technique
4. Kidney biopsy for "marginal" donors (Chapter 2 Fig. 2.4)
5. Expediting the process to the operating room (= shortening CIT)
6. Individual "proactive" intraoperative fluid and drug protocol to decrease delayed function (Tables 6.4-6.8)
7. Targeted early effective immunosuppression (to avoid first time acute rejection) (Figs. 6.2, 6.3, 6.6)
8. Frequent clinic visits in early postdischarge phase, including weekends
9. Proactive problem solving and intervention (i.e., ultrasound, biopsy, cultures)

---

* CIT = cold ischemia time

Table 6.2. Examples of minimum service requirements for a renal transplant program

1. Hospital support system (operating room, anesthesia, administration, outpatient clinic)
2. Organ Procurement Organization efficiency (quality and number of donors, number of organs per donor, hospital development program)
3. Patient referral pattern
4. Waiting list maintenance (optimize medical condition of patients, adequate vascular access, regular clinic visits, communication with dialysis centers)
5. Marketing of program (repeated in-services for dialysis centers, nephrologists, grand rounds for physicians and hospitals, continuing medical education (CME) programs, pamphlets, patient education information)
6. UNOS approval
7. Medicare certification
8. Participation in a patient registry program with a certified Organ Procurement organization
9. Social services provided by a master's prepared social worker
10. Dietetic services provided to assess nutritional needs by the attending physician and a qualified dietician
11. Laboratory services for cross matching of recipient serum and donor lymphocytes for preformed antibodies by an acceptable technique available on a 24 hour basis, to also include phenotyping for donors and recipients, updating and retyping for HLA, screening of recipients

the last examination. It is both practical and courteous to confer with the referring physician regarding the patient's current medical status. Depending on circumstances such as the patient's geographic location, high panel reactive antibodies and other factors likely to predict a positive crossmatch, a back-up patient may be called in and prepared in case the first candidate does not qualify. This practice must be considered carefully due to the cost and emotional impact on the potential recipients. The most important perioperative factors that affect the short-term outcome are the ischemia times and intraoperative blood volume expansion with albumin. Additionally, the intraoperative administration of a calcium antagonist such as verapamil markedly decreases postoperative need for dialysis and improves kidney function. Long-term outcome, on the other hand, is more affected by the type and level of immunosuppression. Early, aggressive immunosuppression with high blood levels of CsA, in the range of 300-400 ng/ml for 16 weeks, results in less than 15% biopsy proven rejection episodes (Figs. 6.2, 6.3, 6.6 and 6.7a-b). The preoperative orders given in Table 6.3 reflect the overall management of the patient prior to a cadaver kidney or pancreas transplant procedure.

INTRAOPERATIVE FLUID AND DRUG PROTOCOLS

ANESTHESIA
General endotracheal anesthesia with an epidural catheter placed for postoperative pain control is preferred. In our experience epidural pain control has mark-

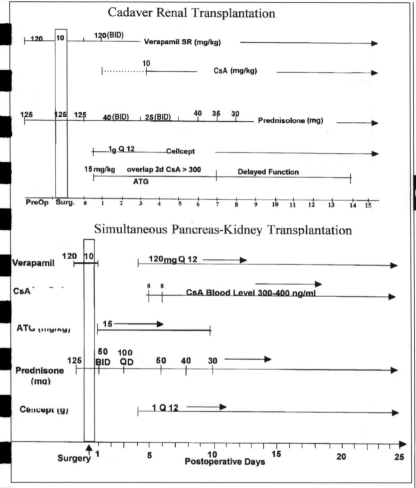

Fig. 6.2. The early postoperative transplant quadruple induction immunosuppression protocol for cadaver kidney (a) and simultaneous kidney/pancreas (SPK) (b) transplants. Currently, the only significant difference between the two protocols is the slightly later initiation of CsA in SPKs.

edly shortened postoperative recovery and improved patient comfort. Ideally the recipient is prepared for surgery by anesthesia and one surgeon, while another team is preparing the kidney or pancreas on the back table. A nasogastric tube is placed in the operating room and removed immediately after extubation or in the recovery room. A Foley catheter is placed (#18-22), and the bladder is irrigated with 500 mg cefazolin in 500 ml of normal saline. The bladder is filled with the

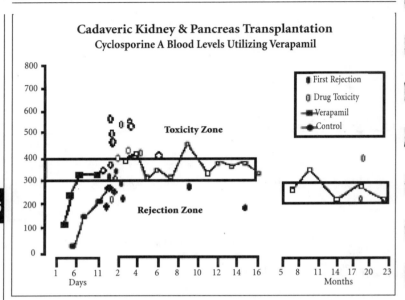

Fig. 6.3. CsA blood levels in relation to acute rejections and nephrotoxicity. The solid line represents mean CsA blood levels in 39 SPK recipients. The boxed area represents a practical target CsA blood level with 8% (3/36) incidence of rejection and 11% (4/36) incidence of CsA nephrotoxicity. (Editor's note: Tacrolimus is interchangeable with CsA; however, the trough level range for this drug in this setting has not been established).The higher CsA levels are induced by concomitant administration of verapamil. This early aggressive but controlled immunosuppression for about 16 weeks also seems to prevent late acute rejection and results in increased long-term graft survival; in fact, no graft losses have occurred in the present study beyond four months.

same solution up to 200 ml, depending on capacity. Care should be taken not to overfill since many patients have small, contracted bladders after years of anuria and hemodialysis. If a small bladder is forcefully expanded, widespread mucosal rupture and bleeding and even bladder rupture may occur with potentially severe sequelae.

Intraoperative management issues are addressed by effective communication with anesthesia. The overall plan, including doses and volumes of drugs and fluids, are reflected in Table 6.4. Special circumstances, such as cardiac or pulmonary problems, must be communicated to the anesthesiologist, with standard protocols adjusted accordingly. A detailed flow sheet or order form is helpful for management, as well as for data and research documentation (Table 6.5). A triple lumen central line catheter is placed, either via percutaneous subclavian or an internal jugular vein route. This line is used for postoperative ATG (Atgam®) infusion and fluid management. In some cases, this line may be discontinued and ATG given through the arterio-venous hemodialysis access.

Table 6.3. Example of preoperative orders for cadaver kidney and pancreas transplant recipients currently used by the author

Cadaver Kidney or Simultaneous Kidney/Pancreas Transplantation
Admission Orders
1. Notify Transplant Resident of patient arrival to hospital.
2. Diagnosis: Cadaver Renal Transplant secondary to _____
3. Notify the Nephrologist on call for medical clearance.
4. Allergies:_____
5. Weigh on arrival, (record in kg) _____
6. Nothing by mouth, except preoperative medications.
7. Record intake and output.
8. Scrub shower with 10% povidone iodine, nipples to knees, then paint with 10% povidone iodine prior to surgery.
9. Send medical emergency Chemistry 18, Amylase, Complete blood count with differential, PT/PTT, Type and Crossmatch for 2 units packed red blood cells (if patient is CMV negative, obtain CMV negative blood.), Hepatitis panel, CMV IgG and IgM.
10. Obtain medical emergency electrocardiogram and chest X-ray, anterior/posterior; lateral views.
11. Obtain operative consent.
12. After medical clearance is obtained,
    a) Start IV with normal saline at 40 cc/hour.
    b) If diabetic, initiate the insulin / dextrose 50 protocol [see Table 6.9]
    c) Atgam skin test applied by MD, read one hour later before administering Solumedrol.
    d) Triple lumen central line to be placed.
    e) Elastic compression stockings and pneumatic compression sleeves to be placed.
    f) Preoperative medications:
        1) Verapamil slow release 120 mg by mouth
        2) Methylprednisolone 125 mg IV.
        3) Ampicillin / Sulbactam (Unasyn) 1.5 g started before skin incision.
        4) If patient is Penicillin allergic, Aztreonam 1g IV, Flagyl 500 mg IV, Vancomycin 1g IV.
13. Obtain final crossmatch results. Attach copy to patient chart on call to OR.

Based on the author's research and clinical experience, an intraoperative infusion albumin dose of 1.0-1.5 g/kg is optimal for maximizing early posttransplant renal function. The albumin infusion is started after anesthesia induction, at approximately 25 g/h in an adult amounting to 75-100 g before kidney reperfusion. A central venous pressure of 15-18 cm H_2O is used as an upper safety guard for prevention of intravascular volume overload. Careful examination of the lungs for left-sided heart failure and for new or louder S3 sounds are also important safeguards against blood volume overexpansion. The goal is to maximize renal blood flow by increasing cardiac output. This is accomplished by slightly overexpanding the blood volume. During the transplant procedure patients lose approximately 1.0 g/kg body weight of albumin. In addition to the albumin infusion, electrolyte solutions are given in a volume of 30-50 ml/kg body weight. This

Table 6.4. Summary of the author's current intraoperative fluid and drug management of renal/pancreas transplant recipients

- Total intraoperative albumin administration of 1.0 to 1.5 g/kg body weight. (SPK recipients may receive 1.5-2.0 g/kg). This amounts to approximately 25 g IV infusions per hour of either 25% or 5% albumin solutions. A CVP of 15-18 cm H_2O is used as an upper limit safeguard at the time of vascular declamping, along with careful determination of fluid status from lung examination.
- Electrolyte (i.e. saline) volume infusions are limited to about 30 ml/kg of body weight. This translates to a slow infusion of 700-1000 cc per hour in a 70 kg man, assuming a 3-5 hour surgery. Volume infusion is adjusted based on dialysis history, "dry weight," actual body weight, and preoperative dialysis fluid shift.
- Mannitol infusions are avoided, since the colloid (albumin) administration overrides the short-acting blood volume expansion effect of mannitol. (Also, mannitol induces histologic changes in the kidney that may be confused with cyclosporine toxicity.)
- Before reperfusion, 125 mg of methylprednisolone is given IV.
- At the time of kidney reperfusion, 80-100 mg of furosemide is given IV. Larger doses have not been proven effective in terms of urine volume increase. In living related transplants, 20 mg increments may be given.
- After vascular clamps are removed, up to 10 mg (2.5 mg increments) of verapamil is injected into the renal artery, observing blood pressure. Blood pressure drop > 10 mm Hg is unusual with concomitant albumin blood volume expansion. If intraoperative hypotension (systolic blood pressure < 100 mg Hg) is present, verapamil administration is not advised.
- In pancreas recipients the somatostatin analog (octreotide acetate, Sandostatin®) 100 mcg is given subcutaneously after induction of anesthesia in cases when enteric exocrine drainage is used.

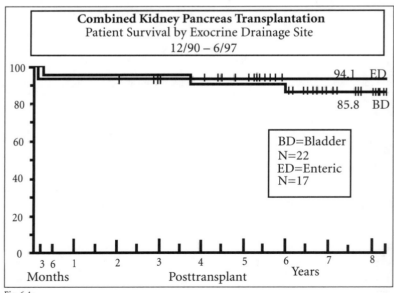

Fig. 6.4a.

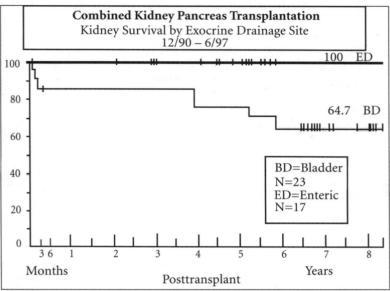

Fig. 6.4b

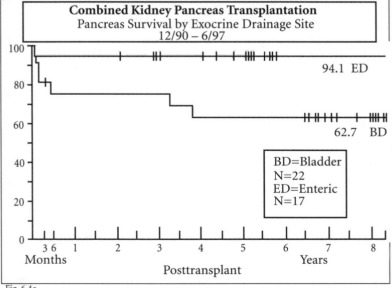

Fig. 6.4c

Fig. 6.4. Actuarial patient (a), kidney (b), and pancreas (c) survival in 39 combined kidney/pancreas transplants as of 7/99, 22 with exocrine drainage to the bladder (BD), and 17 to the small intestine (ED).

Table 6.5. *Example of intraoperative flowsheet/orders used by the author*

Kidney and Kidney/Pancreas Transplantation Surgical/Anesthesia Fluid and Drug Protocol
* For dosing and rates consult with surgeon prior to and during surgery
** Orders pertaining only to pancreas transplant in bold letters

1. **Give Somatostatin (octreotide acetate) 100 mcg subcutaneous injection.**
2. **Low molecular weight Dextran (10%) (Rheomacrodex®) 10 cc/hour.**
3. Give Albumin 25% as follows: Dose: _____ mg/kg; Total dose: _____ g;
   Give: _____ g/hour.
   Aim for central venous pressure at declamping of _____ cm H$_2$O.
4. Give electrolyte solutions as follows: Dose: _____ ml/kg; Total dose: _____ ml;
   Give _____ ml/hour.
5. Methylprednisolone 125 mg before declamping as per surgeon.
6. Furosemide 100 mg before declamping as per surgeon.
7. Verapamil interarterial dose, 10 mg on 30 gauge needle, given by surgeon in operative field.
8. Record anastomosis times: [a] venous anastomosis: (pancreas) _____ to _____;
   (kidney)_____ to _____ ; [b] arterial anastomosis: (pancreas) _____ to _____;
   (kidney)_____ to _____.
9. Record cross-clamp time from procurement surgery : _____.
10. Record time cross-clamp removed in recipient surgery: _____.
11. Calculate cold ischemia time (elapsed) pancreas: _____; kidney:_____.

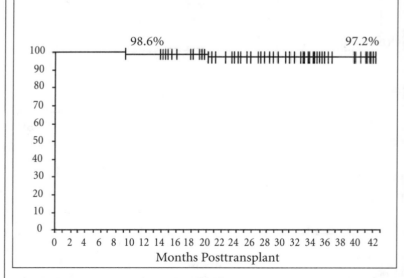

Fig. 6.5a,b. Cadaver renal transplant actuarial patient and graft survival as of 7/99 in SUNY at Buffalo Transplant Program in 74 recipients from 11/95 through 3/98, using the immunosuppression protocols outlined in Figure 6.2a.

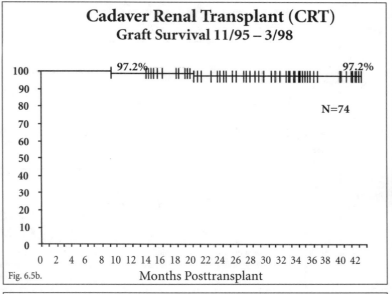

Fig. 6.5b.

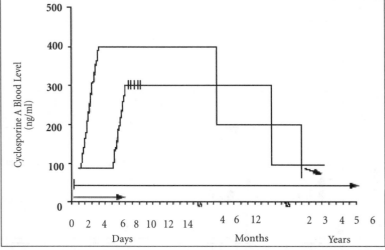

Fig. 6.6. The cyclosporine A blood level target window over time as used by the authors. Atgam 15 mg/kg/ day is started at day 1. CsA is initiated 10 mg/kg in two divided doses when creatinine is < 3 or has fallen 50% from baseline, often within 24-48 h after surgery. CsA levels of >300 ng/ml overlap for 3 days before Atgam is discontinued. A level of 300-400 ng/ml is maintained if tolerated for 14-16 weeks. Failure to tolerate CsA has occurred in 5/74 (6.8%) requiring discontinuation and is more likely in CRT from older donors. HUS occurred in one case, verified FSGN in one case, both of which were started on tacrolimus. Between 6-12 months CsA levels of 200-300 ng/ml, and from 1-2 years 100-200 ng/ml, and beyond 2 years no guidelines exist.

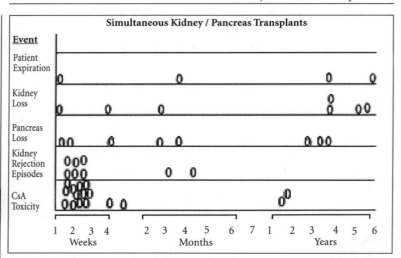

Fig. 6.7a. Posttransplant events (Patient deaths, graft losses, biopsy verified rejection and toxicity episodes) in 39 consecutive kidney/pancreas (SPK) transplants 12/89-6/97 subjected to the immunosuppression protocol outlined in Table 6.10 and Figure 6.2b. There were 21% (8/39) and 36% (14/39) biopsy verified kidney rejections and toxicities respectively within 6 months. The majority of these occurred within 3 weeks after transplantation.

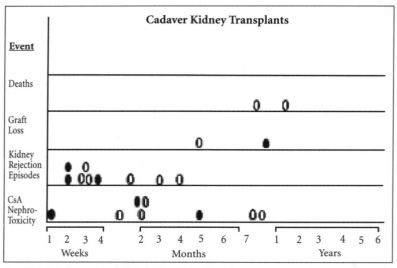

Fig. 6.7b. In 74 subsequent (11/95-3/98) cadaver kidney transplants using a very similar protocol (Fig. 6.2a), the fewer rejection episodes of 12.2% (9/74) may reflect a stricter cyclosporine A blood level target enforcement and the introduction of mycophenolic acid as part of the protocol. Filled circles indicate patients considered high risk pretransplant due to repeat transplant, PRA > 50%, rapid FSGN. The dramatic decrease in biopsy documented nephrotoxicity reflects the lessons learned from the SPK series (Fig. 6.7a), where no rejections occurred at CsA levels above 400 ng/ml (See also Fig. 6.3).

Table 6.6. Effects of intraoperative blood volume expansion with albumin

1) Increased cardiac output with better organ perfusion
2) Early onset of urine output and larger urine volumes
3) Improved renal function (lower serum creatinine, improved glomerular filtration rates and decreased incidence of delayed function)
4) Improved graft survival by 15-20% at one year (lack of delayed function effect)

Table 6.7. Beneficial clinical effects of perioperative verapamil administration in renal transplantation.

- Restoration of renal microcirculation
- Prevention of cyclosporine A (CsA) induced blood flow inhibition
- Induction of elevated CsA blood concentration
- Partial prevention of acute CsA nephrotoxicity
- Reduced posttransplant need for hemodialysis
- Reduced incidence of rejection episodes
- Additive effect when combined with intraoperative blood volume expansion with albumin
- Increased graft survival

Table 6.8a. Beneficial effects of combined albumin (ALB) (1.0-1.5 g/kg) and verapamil (VP) (10 mg intra-arterially) in cadaver renal transplants (Mean ± SD)

	ALB + VP	NONE	P VALUE
Urine output first 24 hrs (L)	5.7 ±3.7	2.2 ±2.1	0.0004
Cr/S at day 7 (mg/100 ml)	2.1 ±1.8	4.3 ±3.5	0.0004
DF	12%	42%	0.0001
Graft survival at 1 year	84%	62%	0.01

Cr/S = Serum creatinine
DF = Delayed function

Table 6.8b. Odd ratio estimates for the decreased risk of delayed function after CRT with perioperative verapamil (VP) and albumin (ALB)

	Decreased Risk of Delayed Function	95% of Confidence Limits	P Value
VP	2.4	1.1–5.0	0.028
ALB	2.3	1.0–5.2	0.018
VP + ALB	5.4	2.9–10.0	0.001

generally translates to 500-1000 ml/h during the transplant procedure, assuming an operating time of 3 h. The volume of electrolyte solution is adjusted to the patient's pretransplant overall fluid status and dialysis history. Furosemide 80-100 mg IV push is given in cadaver transplants immediately prior to reperfusion. For a living related recipient, cautious administration of 20 mg of furosemide is warranted, as high doses may produce large, uncontrollable urine output volumes. Immediately after reperfusion, with hemostasis obtained, an injection of verapamil is given into the renal artery, using a 30 gauge needle. Verapamil is given in 2.5 mg increments over 5-10 min, watching blood pressure changes, up to a total dose of 10 mg. Verapamil is available in 2.5 mg/ml vials of 2 ml (Table 6.4). The injection site is usually the external iliac artery, proximal to the anastomosis, or the aortic patch, while the external iliac artery, distal to the anastomosis, is temporarily clamped using a velcro bulldog clamp. Verapamil makes perfusion of the kidney visibly better and markedly improves the flow, as measured by duplex Doppler ultrasonography. Additionally, intraoperative verapamil decreases the incidence of posttransplant delayed function (need for hemodialysis), and improves glomerular filtration rate. Finally, verapamil partially prevents cyclosporine (CsA) induced vasospasm allowing higher CsA blood levels with less nephrotoxicity.

The intraoperative measures to improve postoperative organ function are summarized in Table 6.4. The effect of intraoperative administration of albumin and verapamil are highlighted in Tables 6.6 and 6.7, respectively. The additive effect of albumin and verapamil on posttransplant outcome is given in Table 6.8 a-b. The authors have abandoned the use of mannitol, a short-acting colloid, which has been replaced by the administration of albumin.

In cases of simultaneous pancreas/kidney (SPK) transplants, the exocrine drainage may be performed using the traditional duodeno-urinary bladder anastomosis or the more controversial intestinal (small bowel) method. The authors prefer the side-to-side duodeno-intestinal anastomosis, which has eliminated the urological complications (Chapter 1, Table 1.7) as well as the almost universal acidosis from urinary bicarbonate losses with the bladder drainage procedure (Chapter 3, Table 3.3). The intestinal drainage avenue has eliminated re-admissions after SPK transplantation in the author's experience.

To avoid overdistention of the donor duodenum as well as the intestinal anastomosis in pancreas transplants, octreotide acetate (Somatostatin®) 100 mcg is given subcutaneously early during the procedure when the enteric drainage procedure is used. The subcutaneous injection of Somatostatin® is continued until the patient is passing gas, eating solid food or has had a bowel movement. The recommended dose is 100 mcg every 12 h. Long-term use of Somatostatin® is not recommended (see also Chapter 3).

PERIOPERATIVE DIABETES MANAGEMENT
Starting preoperatively, the authors prefer to use an insulin and dextrose infusion protocol for perioperative management of patients with insulin dependent diabetes, as outlined in Table 6.9. In cases of central line availability, the 50%

Table 6.9. Insulin/Dextrose infusion protocols currently used by the author

CBG*	D10% Protocol 250 IU in 500 cc NS			D50% Protocol 250 IU in 250 cc NS	
	INSULIN DRIP		D10%	D50%	INSULIN DRIP
	IU*/hr	cc/hr	cc/hr	cc/hr	cc/hr or IU/hr
≤ 70	0.5	1.0	75	17	1
71-100	1.0	2.0	65	17	2
101-150	1.5	3.0	50	17	3
151-200	2.0	4.0	50	17	4
201-250	3.0	6.0	35	17	6
251-300	4.0	8.0	25	17	8
301-400	6.0	12.0	0	17	9
> 400				17	10

* CBG = Capillary blood glucose (mg / 100 ml)
* IU = International Units of insulin

dextrose (D50%) protocol can be used if IV fluid administration is restricted. When peripheral venous IV access is the only choice, the 10% dextrose (D10%) protocol is preferred. These protocols, initially designed by Dr. Philip Raskin at UT Southwestern Medical Center, Dallas, Tx, enable the patient to maintain a blood glucose around 100 mg/dl. Experimental data suggests that maintaining strict blood glucose control early after pancreas transplantation results in better long-term islet cell function.

POSTOPERATIVE ORDERS AND MANAGEMENT

Depending on local hospital traditions, resources and levels of care, patients may go directly to the floor after a brief stay in the recovery room or intensive care unit. An example of postoperative physician orders used by the author is shown in Table 6.10. The authors realize that many centers around the world use markedly different approaches to immunosuppression and patient management depending on research protocols, patient population, philosophy and personal experience and preference. Protocols are used to streamline and expedite peritransplant management and increase safety, and must be updated periodically to reflect medical advances and improved technology.

Most kidney transplant recipients can and should be extubated in the operating room or recovery room. With the fluid and drug protocols utilized in the operating room (Table 6.4), more than 80% of patients will have urine output within 30 minutes, or shortly after skin closure. Initially, the urine may be somewhat blood tinged. Universally, there will be some small clots, often obstructing the Foley catheter. Ideally, with the intravesical implant, the Foley catheter is irrigated from inside before closing the bladder to clear any blood clots.

Table 6.10. *Example of postoperative orders for cadaver kidney and simultaneous kidney/pancreas (SPK) transplants used by the author*

Cadaver Renal/Pancreas Transplant Postoperative Orders
* Orders pertaining only to SPK transplants are printed in bold letters.

1 Admit to postanesthesia recovery area. Condition _____.
   Admit to surgical intensive care unit. Condition _____.
2 Send medical emergency: complete blood count, PT/PTT, chemistry panel (SMA 7), followed by repeat chemistry panel (SMA 7) every 6 hr. Fingerstick glucose every hour to follow Insulin/D50 protocol (see Table 6.9).
3 Give methylprednisolone 125 mg IV at (Time).
4 Vital signs: blood pressure and pulse every 15 min. x 4, every 30 min. x 4, every 1h x 4, every 2 h x 8, then every 4 h: CVP every 4 h, temperature every 2 h for 24 h.
5 Notify MD for temperature > 38°C; pulse rate > 120 or < 60; systolic blood pressure > 160 or < 100; diastolic blood pressure > 110 or < 50.
6 Pneumatic compression stockings to pump.
7 Hourly input and output (I/O). Notify MD of decrease in urine output, or output < 75 cc per hr.
8 Nothing by mouth except ice chips.
   Nothing placed down nasogastric tube.
9 IV Fluids: D5 1/2 NS at 75 cc/hr, as baseline IV to avoid hypernatremia. This volume should be incorporated in the overall IV replacement as outlined in #10.
10 Urine output replacement with normal saline. If urine output > 400 cc/hr call MD. If urine output is 75-400 cc/hr, replace at 100%. If urine output is < 75 cc/hr call MD.
11 Respiratory Care: Turn, cough and deep breathing every 2 hrs. When extubated 40% O$_2$ face mask, as needed. Incentive spirometry every 2h while awake.
12 Weigh on arrival and weigh daily.
13 Foley catheter to dependent drainage. Do not change, remove or irrigate, unless on order from transplant surgeon.
14 Eight (8) h of bedrest, then up to chair as soon as tolerated, at least three times daily. Ambulate as soon as tolerated with assist.
15 Routine labs daily; expanded chemistry panel (SMA 18), complete blood count with differential, cyclosporine levels when patient is taking cyclosporine, magnesium level. Fingerstick glucose every hour to follow Insulin/D50 protocol (Table 6.9).
16 Immunosuppression: postop Day 1
   a. Methylprednisolone 60 mg IV every 12 hrs.
   b. Atgam (15 mg/kg/day) _____ mg IV every day.
   c. CellCept 1 g orally twice daily.
   d. Neoral _____ mg orally twice daily (10 mg/kg/day, or 5 mg/kg/twice daily). Dose modified based on renal function.
   e. Verapamil (Calan) SR 120 mg orally twice daily (Dispense as written).
17 Other medications:
   a. Clonidine patch (Catapres TTS) or IV verapamil if needed for blood pressure control.
   b. Bactrim DS 1 tablet daily.
   c. Acyclovir 400 mg twice daily when creatinine is less than 3.0 mg/dl.*
   d. Unasyn 1.5 g IV every 8 hr. for 24 hr. *
   e. Pain control per epidural catheter per anesthesia.
   f. Somatostatin® (octretide acetate) 100 mcg subcutaneously every 12hrs.
   g. Low molecular weight 10% Dextran (Rheomacrodex®) IV at 10 cc/hr.

* Acyclovir and Unasyn are titrated to posttransplant renal function as follows:
For creatinine clearance < 10 ml/min: Acyclovir 200 mg twice daily.Unasyn 1.5 g twice daily
10-50 ml/min: Acyclovir up to 800 mg three times daily.Unasyn 1.5 g three times daily

Bladder drainage of the pancreas has unique complications, with large bicarbonate losses that must be replaced orally (Chapter 1, Table 1.7 and Chapter 3, Table 3.3). Rather than solutions containing potassium, the authors prefer to use sodium citrate or sodium bicarbonate tablets, since potassium will exacerbate hyperkalemia associated with CsA and tacrolimus. Often, patients have neurogenic bladder dysfunction secondary to diabetes. Urine from distended bladders may reflux into the allograft pancreas and cause pancreatitis, requiring treatment with Foley catheter drainage which usually relieves the symptoms within 12-24 h. Urine leaks through one of the donor duodenal stumps may cause peritonitis and possibly abscess (Chapter 3, Fig. 3.13a). Large leaks require surgical correction. Furthermore, the pancreatic enzymes may be activated in inflamed areas of the urethra, penis or vagina. Bladder drained patients are taught to gently wash their urethral areas with warm water after voiding in order to prevent this complication.

In those patients with excellent renal function, hypomagnesemia may develop, with magnesium supplementation possibly required. Rarely, severe hypomagnesemia may induce hypocalcemia. Some patients develop hypercalcemia because of residual hyperparathyroidism associated with their course of dialysis, which usually does not require treatment. Still others develop significant hypophosphatemia. A potassium containing phosphate supplement should be avoided.

POSTOPERATIVE FLUID MANAGEMENT

A careful history prior to surgery will aid postoperative fluid management. Patients with left ventricular noncompliance will often give a history of shortness of breath, paroxysmal nocturnal dyspnea and dyspnea on exertion after large interdialytic weight gains. Patients with ischemic heart disease may report angina after such weight gains or when large amounts of fluid are removed during hemodialysis. Fluid challenges must be administered with caution in such patients. In addition, knowing the amount of urine produced by the native kidneys will aid the interpretation of the posttransplant urine output.

During the intraoperative period the fluid management is directed at optimizing cardiac filling pressures, resulting in increased allograft renal blood flow and thus, renal function. Excessive or insufficient fluid administration must be avoided. Many transplant recipients have diastolic dysfunction due to hypertension. Excessive fluid administration will result in pulmonary edema. In addition to placing the patient at risk, congestive heart failure will decrease renal blood flow and possibly damage the transplant. In rare situations, pulmonary edema will require hemodialysis with its attendant risk and potential damage to the allograft. On the other hand, insufficient fluid administration will cause inadequate left ventricular filling pressures, decreased cardiac output, suboptimal renal blood flow and an increased likelihood of acute renal failure.

Optimal fluid management requires careful examination of the patient's cardiovascular system every 1-2 h during the initial 24 h after transplant surgery. Frequent examinations are particularly critical if fluids are administered rapidly. These serial examinations must be compared with the patient's preoperative examination. Assessment of the left heart is critical. The lungs must be examined

6

carefully for rales and other signs of left-sided congestive heart failure. Although it is difficult to distinguish atelectasis from failure, skillful patient instruction and proper use of pain medication (ideally epidural pain control) will allow the coughing and deep breathing necessary for elimination of atelectasis. In addition, optimal examination of the lungs requires that the patient be in a sitting position. A new or louder S3 is an important indicator of congestive heart failure. Right ventricular filling pressures may be estimated by examining the neck veins. In most patients, the body habitus will allow accurate assessment of the left and right heart filling pressures without CVP or Swan-Ganz catheters. Should such catheters be used, the wave-forms of the pressure readings must be assessed to be certain that the measurements are accurate.

In contrast to the kidney allograft which receives approximately 20% of the total cardiac output, the pancreas allograft receives only a small portion of the cardiac output, and a recent anastomosis, especially the portal vein, may thrombose if the cardiac output is not optimal.

If the patient is voiding more than 300 cc/hr of urine one may assume adequate renal blood flow. If there is less urine output during the initial 12 h after surgery, an effort should be made to optimize the cardiac filling pressures. If the cardiovascular examination indicates that the patient does not have adequate filling pressures, a bolus of saline (500-1000 ml) and/or albumin (12.5-25 g) is administered over 30-60 minutes. The patient should then be reexamined, as more IV fluids may be necessary. On the other hand, if the examination indicates that the patient has appropriate filling pressures, a large dose of IV furosemide (80-100 mg) may be appropriate. If a patient continues to be oliguric or anuric after these steps, an anatomical problem may be considered. These include obstruction of the ureter or more rarely, a vascular anastomosis problem.

POSTTRANSPLANT HEMODIALYSIS

In about 10-15% of cadaveric transplants there will be delayed function of the renal allograft despite appropriate fluid and drug management. This may occur after prolonged cold ischemia time (> 24 h), or when the allograft is taken from a marginal donor. Even in these circumstances we believe that dialysis should be avoided if possible. Hypotension during posttransplant hemodialysis may damage the renal allograft, and blood dialyzer interactions may exacerbate the acute renal failure of the allograft. Furthermore, dialysis removes indicators of allograft renal dysfunction which may be early clues indicating rejection. Such clues include a rising serum creatinine, volume retention and increasing hypertension.

Careful attention to the fluid, salt and potassium intake may allow dialysis to be avoided. Kayexalate 15-30 g orally or as enemas should be used cautiously, especially in diabetics, because of reports that Kayexalate/sorbitol may rarely be associated with colonic injury.

Absolute indications for posttransplant hemodialysis are uncorrectable hyperkalemia and fluid overload with shortness of breath and pulmonary congestion visible on chest X-ray. If dialysis is deemed necessary, hemodialysis is often preferred to peritoneal dialysis, even in patients with a Tenckhoff catheter in place.

This is particularly true if the peritoneum was invaded during the transplant procedure. Hemodialysis should be performed with a minimum amount of heparin and with a biocompatible membrane. Aggressive ultrafiltration should be avoided if possible to prevent hypotension during dialysis and further injury to the allograft.

EARLY MOBILIZATION

Most patients can take ice chips a few hours after surgery and should be offered a light meal when hungry. The epidural pain control will speed recovery and decrease postoperative constipation from oral pain control medications. Few patients receiving a kidney transplant alone need to be in an intensive care unit, which, in the author's experience, tends to prolong recovery and delay mobilization, as well as increase cost. The pneumatic stockings applied before surgery to prevent deep vein thrombosis should be taken off once the patient is mobile, usually within 24 h after surgery. In patients with adequate bladder capacity the Foley catheter can be removed within 48-72 h if urine output has dropped to less than 200 ml/h and the patient is able to get out of bed by themselves. After the Foley catheter has been removed, furosemide administration should be avoided in the afternoon in order to give the patient more continuous rest during night hours.

DRUGS AFFECTING EARLY KIDNEY FUNCTION

VERAPAMIL

Verapamil slow release (SR 120 mg) is given preoperatively. Also, 10 mg of verapamil is injected in 2.5 mg increments using a 30 gauge needle into the renal aortic patch or into the iliac artery proximal to the anastomosis, while observing blood pressure changes (Table 6.4, and Chapter 2, Fig. 2.16). Verapamil SR 120 mg twice daily is continued orally, with a higher dose if needed for blood pressure control. Verapamil may be contraindicated in patients with some types of heart block. Patients with bradyarrhythmias may benefit from diltiazem in place of verapamil.

ALBUMIN AND VERAPAMIL ARE ADDITIVE

The improved outcome with this perioperative use of verapamil and intraoperative volume expansion with albumin is summarized in Figure 6.3 and Table 6.8a,b. In the authors' experience, when both albumin and verapamil were employed, urine output was 5.7 liters in the first 24 h vs. 2.1 liters when neither drug was used (Table 6.8a). More dramatically, serum creatinine at 7 days was 1.9 mg/dl vs. 4.6 mg/dl and delayed function 12% vs. 42%, respectively. Most importantly, graft survival was 84% at one year when both albumin and verapamil were used vs. 62% when neither regimen was employed (Table 6.8a). When either verapamil or albumin was given alone, intermediate outcomes were obtained. Statistical analysis indicates that the effects of albumin and verapamil are additive, with 2.3 and 2.4 times decreased risk of delayed function respectively when given alone and 5.4 times decreased risk of delayed function when combined (Table 6.8b).

ANGIOTENSIN–CONVERTING ENZYME (ACE) INHIBITORS

Unless the benefits outweigh the risks, ACE-inhibitors, nonsteroidals and beta-blockers should be avoided during the first several months after a renal transplant. The first two agents may impair the glomerular filtration rate, while all of these agents may exacerbate the hyperchloremic hyperkalemic metabolic acidosis associated with CsA and tacrolimus. ACE-inhibitors may be excellent hypotensive agents several months after transplant, when lower doses of CsA or tacrolimus are employed and when a modest increase in serum creatinine may be attributed to the drug rather than to rejection.

THE IMMUNOSUPPRESSION PROTOCOLS

QUADRUPLE DRUG INDUCTION PROTOCOLS

The short-term perioperative overall immunosuppression regimen for kidney alone and for simultaneous pancreas/kidney (SPK) transplants is outlined in Figure 6.2a-b. These protocols are examples of the so-called quadruple induction immunosuppression protocol, in this case using the four drugs: **prednisone, antithymocyte globulin** (Atgam®), **cyclosporine A** (CsA) (Neoral®; SangCya™) and **mycophenolate mofetil** (MMF) (CellCept®). The authors' current early posttransplant immunosuppression protocols are also outlined in Tables 6.10-6.13. While prednisone and Atgam® effectively prevent rejection in the first few days after surgery, CsA and MMF are continued long-term. The authors' overall strategy is early, intense, but controlled immunosuppression for about 4 months, then tapering to lower doses and levels without rejection. With the current formulations of CsA (Neoral®, Sandoz Pharmaceutical Corp., Hanover, NJ and SangCya, SangStat, Menlo Park, CA), CsA levels of 300-400 ng/ml are usually obtained after 1-3 doses. Also see Chapter 12, Table 12.3 for current CsA formulations. Atgam® (Upjohn/Pharmacia, Kalamazoo, MI) is discontinued after three consecutive days overlap with CsA blood levels (TDx/TDxFLx® cyclosporine Monoclonal Whole Blood Assay, Abbott Labs, Abbott Park, IL) of at least 300 ng/ml. Although methods of monitoring CsA levels will vary between centers, the important concept is to maintain Atgam® therapy until adequate levels of CsA are achieved allowing an acceptable overlap.

Methylprednisolone is given in three divided doses of 125 mg over the first 24 h, the first dose given as premedication, the second during surgery prior to the release of vascular clamps and the third dose approximately 8 h after surgery (Fig. 6.2 a-b). Prednisone dose is then tapered over time, but not eliminated (Table 6.13).

Antithymocyte globulin (Atgam®) is started within 6 h after surgery at a dose of 15 mg/kg of body weight, diluted in 500 ml D5W or saline and infused over

---

Table 6.11. Atgam induction infusion protocol used by the author

---

1. Atgam (ATG) skin test must be administered by MD on admission for transplant. Skin test is read 1 hour after administration and result documented in progress note. A flare greater than 10 mm precludes the use of Atgam.
2. ATG dose: 15 mg/kg dry weight rounded downward to nearest 250 mg. _____ mg
3. Premedication 30 minutes prior to infusion
   First dose _____ date:
   Solumedrol 125 mg IV
   Tylenol 650 mg as needed, orally
   Benadryl 50 mg oral or IV
   Three subsequent doses:
   Tylenol 650 mg orally
   Benadryl 50 mg IV
4. Infuse _____ mg ATG in 500 cc D5NS over 4 h through central line.
   Dialysis A/V access may be used for infusion with order from transplant attending.
5. Vital signs:
   a. Dose #1 and #2:
      1. prior to infusion
      2. every 15 min. x 4
      3. every 1 hour x 4
      4. termination of infusion
   b. All subsequent doses:
      1. prior to infusion
      2. every 30 min. x 2
      3. every 1 hour x 4
      4. termination of infusion
6. Notify MD for:
   _____ >BP >_____ mm/Hg or _____ < HR < _____ /bpm
7. If patient develops shortness of breath, wheezing or skin rash discontinue infusion immediately and notify MD immediately.
8. Patient to remain on transplant unit for entire infusion. Transplant resident must be on transplant unit during first 2 h of first infusion and in hospital for remainder of first infusion.

4-6 h (Table 6.11). An Atgam® skin test is done preoperatively and read before administration of the first dose of methylprednisolone. In the rare case of a positive (> 10 mm flare) test suggestive of potential sensitivity or anaphylactic reaction to Atgam®, OKT3® (Ortho Biotech, Raritan, NJ) 5 mg IV daily is substituted. If OKT3® is used, a more conservative fluid regimen is necessary, as dictated by the manufacturer. The first dose is preferably given during surgery, which may alleviate the common first dose reaction of fever, chills and muscle aches. The manufacturer also restricts the use of OKT3® in cases of fluid overload (Editor's note: two new immunosuppressive medications, Daclizumab or Basiliximab may, in the near future, replace current induction agents. See Chapter 12, Table 12.6).

The Atgam® infusion orders are outlined in Table 6.11. The daily dose may be adjusted downward for low white cell count (< 4000 cells/mm³) or platelet (< 100,000 cells/mm³) count. A more accurate tool to guide Atgam® or OKT3® dosing is badly needed. Thymoglobulin (SangStat, Menlo Park, CA) is FDA approved for rejection, but is currently being investigated as an induction agent.

Table 6.12. Example of Orthoclone OKT3 rejection administration orders used by the author

Prior to initiation of OKT3 therapy for rejection the following are recommended:
Renal biopsy; chest X-ray within 24 h free from infiltrates and signs of fluid overload;
weight gain of less than 3% over 7 days.
Administration:
 • Administer via heparin trap:
    1 mg/kg Solumedrol as premedication with first 1-3 doses
    5 mg OKT3 IV bolus over 5-10 minutes
 • Repeat administration procedure for 7-10 days.
 • Maintenance immunosuppression dosages should be administered at the discretion of
   the attending physician.
 • Vital signs as follows for doses 1 and 2:
    every 15 min. x 2 h
    every 30 min. x 2 h
    every hour x 2 h
    then every 4 h
 • Tylenol 650 mg every 4 h as needed for fever > 101°F (39°C).
 • Lomotil 1-2 tablets four times daily as needed for loose stool, not to exceed 8 tablets
   per day.
 • Compazine 10 mg intramuscularly or orally as needed for nausea, vomiting.
 • Benadryl 50 mg orally every 4-6 h as needed
 • Daily complete blood cell count(WBC with differential, platelets), serum electrolytes.

Table 6.13. Methylprednisolone/Prednisone dosing protocol used by the author

Time	Dose	Dose/24 h (mg)
	Methylprednisolone (Iv)	
Day of Surgery	125 mg on call	375
	125 mg before declamp	
	125 mg 8 hrs after second dose	
Postop Day 1	60 mg every 12 hrs	120
	Prednisone (Oral)	
Postop Day 2	40 mg twice daily	80
Postop Day 3	30 mg twice daily	60
Postop Day 4	25 mg twice daily	50
Postop Day 5	45 mg once daily	45
Postop Day 6	40 mg once daily	40
Postop Day 7	35 mg once daily	35
Postop Day 8-15	30 mg once daily	30
Postop Day 16-30	25 mg once daily	25
Postop Day 31-60	20 mg once daily	20
Postop Day 61-90	17.5 mg once daily	17.5
Postop 6 Months	15 mg once daily*	15
Postop 9 Months	12.5 mg once daily	12.5
Postop 12 Months	10 mg once daily	10
Postop 12-24 Months	5-10 mg once daily	5-10
Long-term (> 24 Months)	5 mg daily	5

*Individual patient steroid taper may vary dependent upon the history of rejection
episodes and/or steroid side effects.

CsA (Neoral® or SangCya™) administration is initiated when there is adequate renal clearance, as evidenced by a marked drop in serum creatinine (Cr/s) (i.e., 2 mg/dl) over 24 h or when Cr/s is below 3.0 mg% or decreased to 50% of preoperative baseline level. The initial dose of CsA is 8-10 mg/kg body weight. In cases of delayed function, (need for dialysis) Atgam® may be continued, perhaps at reduced dose or every second day for up to 21 days while CsA is introduced at lower doses as tolerated by the recovering kidney function (Fig. 6.2). Dosing is adjusted for CsA blood 12-hour trough levels of at least 300 ng/ml. It is critical that the CsA levels reflect accurate 12-hour trough readings. A CsA blood trough level between 300-400 ng/ml is maintained for about 16 weeks, followed by a target level of 200-300 ng/ml for up to 12 months (Figs. 6.2, 6.3 and 6.6). Occasionally, lower levels are required because of clinical nephrotoxicity from CsA or even discontinuation due to development of hemolytic uremic syndrome (HUS).

The fourth drug is **mycophenolate mofetil** (CellCept®, Roche Labs, Nutley, NJ) which, in many programs, has replaced azathioprine (Imuran®, Glaxo Wellcome, Research Triangle Park, NC). The standard dose of CellCept® is 1.0 g twice daily, which can be increased to a total of 3 g daily. However, more often the dose has to be decreased due to gastrointestinal symptoms, most commonly diarrhea.

CYCLOSPORINE A BLOOD TROUGH LEVELS

Our early experience with higher CsA levels in simultaneous pancreas/kidney (SPK) recipients is illustrated in Figure 6.3. The solid line represents the mean CsA blood levels in 36 consecutive SPK recipients at The Parkland Memorial Hospital in Dallas followed for 24-60 months. The boxed area represents the target CsA blood levels as outlined by the protocol. The overall incidence of kidney biopsy verified rejections within six months was 22% (8/36). In fact, all acute rejections occurred within 16 weeks (Fig. 6.3). There were no rejections at levels above 400 ng/ml, two at levels between 300-400 ng/ml and six at levels below 300 ng/ml. In contrast, of 14 biopsies diagnosed with renal toxicity within 16 weeks, 10 occurred at CsA levels above 400 ng/ml. Although the use of higher CsA blood levels has been controversial, it is the authors' bias that the associated risk of CsA nephrotoxicity is small compared to the markedly increased graft loss associated with the treatment of acute rejections. In the 36 SPK recipients a Cox regression statistical analysis with time dependent measurements of CsA levels has predicted a significant decrease in the incidence of rejection with higher CsA blood levels ($p > 0.04$, with a relative risk of 0.57/100 ng/ml of CsA), meaning that the risk of rejection is nearly halved for each 100 ng/ml increment in CsA blood levels in the early transplant period. The perioperative use of verapamil partially protects the kidney from CsA toxicity, thereby allowing higher CsA levels. Maintenance of CsA levels between 200-300 ng/ml for up to 2 years in this SPK series has been associated with two instances of deteriorating renal function requiring renal biopsies, both of which were diagnosed as CsA toxicity. Accordingly, in this series of 36 SPK transplants from Parkland Memorial Hospital and 3 SPK from SUNY, Buffalo, there have been no rejections between 6-24 months (Fig. 6.4). The incidence of early rejection episodes within 16 weeks was 21% (8/39), with no rejections at the

Table 6.14. Causes and time of pancreas graft loss within the first year posttransplant according to exocrine drainage modality

	Bladder n = 22	Enteric n = 17
Death	1 (4m*)	1 (7d*)
Thrombosis	3 (8,11,14d)	1 (7d)
Rejection	1 (2m)	0
Abscess	1 (4m)	0
Total	6 (27%)	2 (12%)

* m = months, d = days

present time in 32 patients followed beyond 6 months up to 24 months (Fig. 6.3). Causes of pancreas graft losses are given in Table 6.14 and long-term patient, pancreas and kidney graft survival are shown in Figure 6.4a-c.

Subsequently, using a very similar protocol in 74 consecutive cadaver kidney recipients at the transplant program at SUNY at Buffalo, NY, there have been 9 (12.1%) episodes of biopsy verified rejection and 8 (10.8%) patients with CsA nephrotoxicity, with 2 (2.7%) graft losses (Fig. 6.5). Among these recipients, 5 were considered high risk pretransplant due to repeat transplant, PRA > 50% or rapid FSGN, and accounted for 3 rejection episodes, 3 instances of CsA toxicity and 1 graft loss (Fig. 6.7b). When these patients are removed from the calculation, the remainder of the patient cohort shows 8.5% (6/71) rejection episodes and 7% (5/71) CsA toxicity. The CsA levels and treatments for these are detailed in Table 6.15.

POSTTRANSPLANT RENAL ALLOGRAFT DYSFUNCTION

Hypovolemia

While acute rejection is treatable when the diagnosis is made expeditiously, most patients with increasing serum creatinine after transplant have another diagnosis. Hypovolemia is a common problem in patients with bladder drainage of the allograft pancreas. The pancreas secretes large quantities of essentially isotonic fluid into the bladder, and often these losses are not replaced. Patients with end-stage renal failure on hemodialysis are taught to restrict their intake of both free water and sodium. Volume repletion depends upon salt intake. Patients should not be told simply to drink more water because a hypovolemic patient ingesting water but no salt will have high levels of ADH and become severely hyponatremic. Instead, the patients must be instructed to ingest more sodium and to drink according to their taste. It is our practice to prescribe salt as medication for hypovolemic patients, as one level teaspoon of table salt contains approximately 2 g of sodium. This may be sprinkled on food or consumed as commercial bouillon, where one package contains approximately 1 g of sodium. Thus, drinking four cups of bouillon is approximately the same as giving one liter of normal saline

Table 6.15. Renal biopsy statistics SUNY at Buffalo Transplant Programs 1996-98

Patient Level	CsA* Wks	BX-	Cellular	Vascular	Toxicity	Other	Treatment	High**
1	277	11	n	n	n	normal	none	
2	370	2	n	mild	n		OKT3,PH, SM	y
3	384	1	n	n	n	HUS	d/c CsA,PH	
3	89	2	mild	mild	n		SM	
4	429	8	n	n	y		lower CsA dose	
5	324	6	n	n	y		lower CsA dose	
6	362	8	mild	mild	n		SM	
7	288	3	mild	n	n		SM	
8	175	32	n	n	y		reduce CsA	
9	130	8	n	n	y		reduce,d/c CsA	
9	0	12	mild	n	n		OKT3 + FK	
10	388	1	n	n	y		reduce, d/c CsA	y
10	0	3	n	n	n	FSGN	PH + FK	
11	270	6	mild	n	n		SM	
12	369	2	mild	n	n		none	y
12	190	20	n	n	y		reduceCsA	
13	193	40	n	n	y		reduce CsA	
14	171	2	mild	n	n		OKT3,SM	
15	FK4.2	8	n	n	n	HUS	PH	y
16	419	4	mild	n	n		OKT3,SM	y
16	143	8	n	n	y		reduce,d/c cya + FK	

HUS = hemolytic uremic syndrome   PH=plasmapheresis
FSGN = focal sclerosing glomerular nephritis   CsA=cyclosporine
SM = solumedrol   FK=FK506
* = TDx/TDxFLx® CsA Monoclonal Whole Blood Assay, Abbott Labs, Abbott Park, IL
** = High pretransplant immunologic responder, i.e. PRA >50%, retransplant, rapid
FSGN. In this population, N(high responder) = 5, of whom 3 experienced biopsy proven
rejection.

intravenously. All package labels should be read carefully to be certain that the product does not contain potassium.

Severe hypovolemia may also occur in some patients with kidney only transplants. Hypovolemia may also be a problem in patients who have diarrhea. In some cases the diarrhea may be a side effect of drugs, i.e., tacrolimus and mycophenolate mofetil, but the possibility of CMV enteritis should also be considered.

Another common cause of renal allograft dysfunction is overly aggressive treatment of hypertension or use of diuretics. Anatomic problems, especially lymphoceles, should also be considered and diagnosed with ultrasound studies of the allograft (see Chapter 10). CsA and tacrolimus nephrotoxicity is discussed below.

TREATMENT OF REJECTION

Suspected rejections should be verified by tissue diagnosis. A needle biopsy of the kidney is required to confirm a diagnosis of rejection, CsA toxicity or other causes of renal dysfunction including recurrent disease. A needle biopsy should be performed under ultrasound or CT guidance (Chapter 10). There is significant morbidity associated with high doses of immunosuppressive drugs given as rejection therapy, including infection and malignancy. Therefore, patients suspicious for rejection should not receive anti-rejection medications until a tissue diagnosis has been made.

As is the case with the immunosuppression protocols, rejection therapy strategies vary between transplant centers. The following summarizes the author's current protocols.

Acute cellular rejection

Acute cellular rejection is treated with methylprednisolone 1000 mg in 250-500 ml of saline infused over 10-30 minutes. This is given in four doses on days 1, 2, 3 and 5, while baseline immunosuppression is maintained or adjusted for therapeutic CsA blood levels.

Acute cellular rejection which does not respond to therapy with methylprednisolone, defined as lack of creatinine stabilization within 3-4 days, is considered for OKT3® therapy (Table 6.12). Patients who have not had Atgam® as part of immunosuppression induction could receive it at this point (Table 6.11). OKT3®, Atgam® or Thymoglobulin® (SangStat, Menlo Park, CA) for rejection therapy are given for 7-14 days pending clinical response. For rejection therapy, Atgam® and Thymoglobulin® infusion require central line placement, unless a working vascular access is present. OKT3® may be given through a peripheral vein, making it suitable for the outpatient setting.

Acute vascular rejection

Acute vascular rejection is usually resistant to steroid therapy. The authors currently treat this with OKT3® for 7-10 days. Concomitant plasmapheresis treatments may be attempted. There is currently no agreed upon plasmapheresis protocol or strong scientific data in support of its efficacy.

Chronic rejection

Chronic rejection is called chronic allograft nephropathy by many authors, and although indistinguishable, actually has two major components: one alloantigen-dependent and the other alloantigen-independent. The latter is a maladaptive response of the kidney to nephron loss. There is no widely accepted therapy of the alloantigen-dependent component of chronic allograft nephropathy. Some centers add mycophenolate mofetil (CellCept®) and decrease the CsA dose in these patients to alleviate possible CsA toxicity. The alloantigen-independent component should be managed with aggressive control of hypertension (especially with ACE-inhibitors), hyperlipidemia and hyperglycemia in patients with diabetes. Therefore, chronic allograft nephropathy may represent several other processes, such as recurrent disease, rapidly progressing atherosclerosis, drug toxicity or a combination of these. When there are concomitant acute rejection changes superimposed, often reflected by a rapid increase in serum creatinine, pulse steroids may be attempted in addition to pharmacologic adjustments.

CYCLOSPORINE A (CsA) NEPHROTOXICITY

CsA nephrotoxicity (Fig. 6.3), is more common at high CsA blood levels. When diagnosed by biopsy, the dose of CsA is decreased slowly, usually with a corresponding fall in serum creatinine. Acute CsA toxicity is a reversible injury. The authors much prefer CsA toxicity to acute rejection. Repeat rejections leading to chronic allograft dysfunction is the major cause of late graft loss. It is the authors' experience that early high CsA levels prevent acute as well as chronic rejection, resulting in long-term graft survival in excess of 90% (Figs. 6.3, 6.4a-c, 6.5).

Cyclosporine A (CsA) (Neoral®; SangCya™) Dosing Strategy

CsA trough (12 or 24 hour) blood levels are used to guide CsA dosing. Some centers use a more elaborate determination of several CsA levels delineating the so-called area under the curve (AUC). Repeated low (less than 300 ng/ml within 12-14 weeks posttransplant) or high (> 400 ng/ml) CsA blood levels are corrected by gradually increasing or decreasing the CsA dose by 25-50 mg daily. Rapid daily changes in CsA dosing produce unpredictable blood levels and jeopardize patient management. Occasional high or low CsA levels with a stable dose and unchanged kidney function should be watched and not acted upon. In the authors' experience with 74 CRT and 39 SPK patients using this protocol, proper CsA dosing requires reacting to several days' trend in CsA levels rather than individual numbers. Early after transplantation, CsA levels of 400 ng/ml and above are maintained with consideration of patient presentation, including stable renal function and decreasing serum creatinine values.

6

THE IMMUNOSUPPRESSION PROTOCOL TIMELINE

SHORT-TERM IMMUNOSUPPRESSION: (0-6 MONTHS)

The protocols for immunosuppression outlined thus far have been concerned with the perioperative and early postoperative periods after transplantation (Figs. 6.2, 6.3 and Table 6.10). This postoperative time to 4-6 months is the most important period, since early acute rejection episodes are the main determinant in the future development of acute, repeat and chronic rejections which result in deteriorating function and graft loss.

With the increasing success of solid organ transplantation long-term outcome, such as 5 and 10 year graft survival, has become the focus of outcome goals. As increasing numbers of patients have many years of good transplant function, chronic immunosuppression is of great concern. There are three concomitant issues in the long-term: 1) continued good graft function, without rejection, 2) balanced against drug toxicity and the 3) risks of long-term systemic immunosuppression, mainly infection and malignancy.

Perioperative Immunosuppression (0-14 days)

The purpose of immediate perioperative immunosuppression up to about 14 days is to allow the kidney to recover from procurement events (mainly ischemia), while the drug combination protocol avoids nephrotoxicity yet ensures adequate immunosuppression. The authors have elected to use the quadruple induction immunosuppression protocol, (Fig. 6.2a-b and Table 6.10) avoiding CsA nephrotoxicity while the kidney is regaining adequate function. In cases of excellent early renal function usually reflecting an uncomplicated donor and procurement situation, CsA can be introduced as early as day 1 or 2 posttransplant. Some programs use intravenous CsA starting immediately after surgery at a dose of about 3-4 mg/kg. Infusions are provided on a 24 hour basis. The suggested IV dose is about a third of the oral dose (Chapter 12, Table 12.3).

The Early Postoperative Period (2 weeks to 4-6 months)

The second phase of postoperative immunosuppression begins after the induction period and continues to approximately six months as outlined in Figures 6.3 and 6.6. The overall goal in this immunosuppressive phase is to avoid the occurrence of acute rejection episodes while minimizing CsA or tacrolimus nephrotoxicity and avoiding opportunistic infections from generalized overimmunosuppression. In this time period, the authors have elected to maintain a CsA blood target level of 300-400 ng/ml while also giving mycophenolate mofetil in a dose of 1 gram twice daily, and prednisone in a tapering dose (Table 6.13). These CsA levels are approximately twice as high as the authors have used previously. Simultaneous administration of a calcium antagonist such as verapamil or diltiazem partially protects the kidney from CsA induced vasospasm. The effects of perioperative verapamil administration are outlined in Tables 6.7 and 6.8.

LONG-TERM IMMUNOSUPPRESSION (6 MONTHS AND BEYOND)

If the reader finds the short-term immunosuppressive protocols confusing, the long-term guidelines, lacking strong scientific backing, are even more vague. Future randomized prospective clinical trials and experimental animal research will further refine the protocols in a variety of clinical settings. The following outline provides one example of a long-term management style. Rather than religiously adhering to a specific protocol, the reader should be aware of the issues and select protocols that fit specific local circumstances. The issues addressed below are the doses and blood levels of CsA, mycophenolate mofetil and prednisone respectively, over time.

Cyclosporine A (CsA) Blood Levels (6-12 months)

CsA whole blood levels between 6 and 12 months (Figs. 6.3 and 6.6) are currently targeted for 200-300 ng/ml, or a 100 ng/ml decrease from the early posttransplant period (Fig. 6.3). This CsA blood level therapeutic window is based on the fact that no rejections have occurred in 36 simultaneous kidney/pancreas transplants and 74 cadaver kidney transplants followed for up to 48 months at this target blood level. The validity of this protocol is further supported by the only 12% incidence of rejection episodes in 74 consecutive cadaver kidney transplants. Of these 113 patients (74 CRTs and 39 SPK) followed for more than one year with only 4 instances (3.5%) of biopsy-verified, clinically significant nephrotoxicity between 6-24 months. In contrast, of the 20 cases of toxicity early after transplantation 10 (50%) occurred at blood levels above 400 ng/ml. In other words, the incidence of rejection and toxicity events sharply decline as time elapses after transplant (Figs. 6.3, 6.7a,b).

Cyclosporine A (CsA) Blood Levels (12-24 months)

CsA whole blood levels beyond 12-24 months are shown in Figure 6.3. As we move further along the timeline from the transplant surgery, the blood CsA level required becomes more obscure, especially if combined with other drugs such as azathioprine or mycophenolate mofetil (MMF). Sirolimus (Rapamune®), the most recent addition to the immunosuppressive drug arsenal, may prove quite effective in this regard, namely eliminating rejection with lower CsA blood levels. Furthermore, new effective drugs including MMF, tacrolimus and sirolimus lead to the question of which drug, if any, should be the first line of immunosuppression. At this time, the authors have chosen to target a CsA blood level of 100-200 ng/ml for 12-24 months, while maintaining MMF at 1 gram twice daily, if tolerated, and prednisone at low dose. For reasons mentioned above, this strategy is likely to change in the near future.

Immunosuppression Protocol Timeline Beyond 24 Months

The guidelines for choosing drugs and dosing strategy for long-term immunosuppression are not well defined, but nonetheless are of paramount importance for the patient. The protocols range from maintaining triple drug immunosuppression (CsA, MMF and prednisone) to the issue of tolerance, or recognition of the organ as self, implying no immunosuppression is needed. Research into the issue of tolerance is ongoing, but as yet there is no specific test or indicator as to in whom and when this might occur. The authors' current policy is to maintain low level immunosuppression in a patient who has normal and stable function and no previous rejection episodes. This usually means a prednisone dose of no less than 5 mg daily, CsA blood levels around or less than 100 ng/ml and azathioprine dose of 50-75 mg daily or MMF dose of 250-500 mg twice daily.

DECREASING IMMUNOSUPPRESSION OVER TIME

When decreasing immunosuppression, the authors have selected to first decrease prednisone due to its complication profile. At a daily dose of 5 mg, the negative effect on bone metabolism is minimal (Table 6.13, Chapter 12, Table 12.4). Complete discontinuation of prednisone may not be wise because of the risk for acute or chronic rejection. Whether or not to further decrease other immunosuppressive drugs such as azathioprine, MMF or CsA is an open question.

Specific Strategies

A program's immunosuppressive strategy is a reflection of many factors. Long-term immunosuppression should be adjusted to the individual patient's clinical course. In cases of progressive impairment of kidney function with a biopsy showing chronic changes without (acute) cellular infiltrates, the authors choose to decrease or discontinue CsA or tacrolimus to eliminate drug nephrotoxic effect. Also, in such patients azathioprine may be replaced with MMF. There is some data suggesting transplant arteriopathy may be positively affected by MMF.

The future of immunosuppression protocols is likely to be even more complex as more immunosuppressive drugs become available with different actions and side effect profiles (see Chapter 12).

SELECTED READING

1   Conger J. Hemodynamic factors in acute renal failure. Adv Renal Repl Ther 1997; 402 (Suppl 1):25-37.
2   Davidson IJA, Munschauer CE. Albumin, the natural colloid: Experimental data and clinical implications. In: Haljamae H, ed. Bailliere's Clinical Anaesthesiology. London: Bailliere Tindall, 1997; 11(1):81-104.
3   Dawidson I, Ar'Rajab A, Dickerman R et al. Perioperative albumin and verapamil improve early outcome after cadaver renal transplantation. Trans Proc 1994; 26(6):3100-01.
4   Dawidson IJA, Ar'Rajab A. Perioperative fluid and drug therapy during cadaver kidney transplantation. In: Terasaki PI and Cecka JM, eds. Clinical Transplants. Los Angeles: UCLA Tissue Typing Laboratory 1992:267-282.
5   Dawidson IJA, Berglin E, Brynger H, Reisch J. Intravascular volumes and colloid dynamics in relation to fluid management in living related kidney donors and recipients. Crit Care Med 1987; 15(7):631.
6   Dawidson, IJA, Sandor SF, Coorpender L et al. Intraoperative albumin administration affects the outcome of cadaver renal transplantation. Transplantation 1992; 53(4):774-782.
7   Dy G, Raza R, Mendez M. The clinical biochemical effect of calcium channel blockers (CCB) in organ transplant recipients (TR) on cyclosporine (CsA). Trans Proc 1991; 23:1258.
8   Fry WR, Dawidson IJA, Alway CC, Rooth P. Cyclosporine A induces decreased blood flow in cadaveric kidney transplant. Trans Proc 1988; 20(3):222.
9   Hakim RM, Wingard RL, Parker RA. Effect of the dialysis membrane in the treatment of patients with acute renal failure. N Eng J Med 1994; 331:1338-1342.
10 Halloran PF, Homik J, Goes N et al. The "injury response": A concept linking nonspecific injury, acute rejection and long-term transplant outcomes. Trans Proc 1997; 29:79-81.
11 Hauser AC, Derfler K, Stockenhuber F et al. Effect of calcium channel blockers on renal function in renal-graft recipients treated with cyclosporine. N Eng J Med 1991; 324:1517.
12 Kahan BD, Welsh M, Rutzky LP. Challenges of cyclosporine therapy: The role of therapeutic monitoring by area under the curve monitoring. Therapeutic Drug Monitoring 1995; 17(6):621-4.
13 Loutzenheiser R, Epstein M. The renal hemodynamic effects of calcium antagonists. In: Epstein M, ed. Calcium Antagonists and the Kidney. Philadelphia: Hanley & Belfus 1990:33.
14 Lu C. Ischemia, injury and renal allograft rejection. Cur Op Nephrol Htn 1996; 5:107-110.
15 Palmer B, Dawidson IJA, Sagalowsky A et al. Calcium channel blockers improve the outcome of cadaver renal transplantation. J Am Soc Nephrol 1990; 1:767.
16 Shoemaker WC, Appel P, Kram HB et al. Prospective trial of supranormal values of survivors as therapeutic goals in high risk surgical patients. Chest 1988; 94:1176.
17 Troppmann C, Gillingham KJ, Benedetti E et al. Delayed graft function, acute rejection and outcome after cadaver renal transplantation. Transplantation 1995; 59:962-968.
18 Tullius SG, Tilney NL. Both alloantigen-dependent and independent factors influence chronic allograft rejection. Transplantation 1995; 59:313-318.
19 Gaber AO, First MR, Tesi RJ, Gaston RS et al. Results of the double-blind, multicenter, phase III clinical trial of Thymoglobulin versus Atgam in the treatment of acute graft rejection episodes after renal transplantation. Transplantation 1998; 66:29-37.

Histocompatibility Testing in Organ Transplantation: General Policies and Procedures

Thomas Shanahan

7

The purpose of histocompatibility testing is to optimize graft survival in transplant recipients by selecting the most suitable donors and avoiding those combinations that will result in unfavorable outcomes. These goals are generally accomplished in three ways: by determining the human leukocyte antigen (HLA) profiles of the donors and recipients, by measuring HLA-associated antibody reactions against reference cell panels, and by assessing donor-specific humoral immune responses in the histocompatibility crossmatch assay.

HLA TYPING

APPLICATIONS
Cadaveric Transplantation
HLA antigen matching remains the primary criteria for the allocation of cadaveric kidneys. Because of stringent time constraints, most heart, lung and liver programs place less emphasis on HLA matching. However, the cadaver antigen profile is an important consideration for all HLA sensitized patients, particularly when it becomes necessary to avoid specific recall or previous mismatched antigens.
Living Related Transplantation
HLA phenotyping determines the degree of antigen identity between patients and their living-related donors. These relationships are expressed as identity, nonidentity, or haploidentity when all, none or half of the antigen profiles are shared respectively.

INDICATIONS

The HLA phenotypes of potential graft recipients should be determined at the time of initial evaluation. The cadaver typing is performed as part of the donor evaluation process. For living-related donations, HLA phenotypes should be performed in advance of transplantation in order to assess the degree of HLA identity between the donor and recipient.

METHODS

HLA Serology

The serological typing method detects HLA antigens at the cell surface. The antisera used must permit the identification of all major class I (HLA-A and HLA-B) and class II (HLA-DR) antigens as recognized by the World Health Organization (WHO). Attempts should be made to identify the public (broad) and definable split (private) antigens associated with each HLA allele.

DNA Testing

The DNA based molecular approaches for the assignment of HLA phenotypes permit the identification of HLA-associated genes independent of antigen expression. These methods measure DNA amplifications in response to various HLA-specific growth initiators (primers) or detect the presence of HLA gene sequences through the use of specific detectors (probes).

SPECIMEN REQUIREMENTS

Serologic Typing of Recipients and Living Related Donors

HLA typing of these individuals generally requires 30 ml of heparinized (green top vacutainer tubes with sodium heparin) or ACD anticoagulated (yellow top vacutainer tube) whole blood.

Serologic Typing of Cadaver Donors

Because of the adverse effects of steroids and blood transfusion on HLA antigen expression, peripheral blood is not the ideal specimen in most cadaver donor situations. Instead, preharvest lymph nodes are the preferred specimens. A section of spleen obtained at the time of harvest is also desirable for posttransplant follow-up testing. Donor tissue specimens must not be placed in a fixative. Instead, they should be placed in a nutritional transport medium such as RPMI or MEM.

DNA Typing

For many DNA-based typing procedures, 3 ml of EDTA (lavender top vacutainer type) or ACD (yellow top vacutainer tube) anticoagulated whole blood is sufficient. However, specific requirements may vary according to method and manufacturer. Procedures are also available for the extraction of DNA from tissue samples.

INTERPRETATION

HLA alleles or antigens are assigned by the reactivity patterns of specific antisera in the serologic assay or the detection of specific DNA by the molecular assays. The objective is to detect two alleles for each genetic locus (HLA-A, HLA-B and HLA-DR). If only one allele is identified, the second is designated as a blank allele. This usually indicates genetic homozygosity at that particular locus or the expression of an allele not yet identifiable by the current technology.

HLA ANTIBODY SCREENING

APPLICATIONS
The primary purpose of antibody screening is to detect alloreactive antibodies directed against HLA antigens. Whenever possible, the precise antigenic specificity should be determined in order to identify unacceptable, high-risk donor antigens. HLA antibody screens also serve to determine the extent of sensitization, thereby acting as indirect predictors of crossmatch compatibility.

INDICATIONS
All potential recipients awaiting transplantation should be tested for the presence of HLA-associated, panel reactive antibodies (PRA) on a regular basis. If the absence of antibodies (PRA = 0%) is consistent or maximum sensitization (PRA >80%) is confirmed, testing may be performed at less frequent intervals. However, additional sensitizing events, such a blood transfusions, necessitate retesting.

METHODS
Standard Panel Testing
The screening of recipient sera against freshly prepared panel cells involves the collection of lymphocytes from individuals of predetermined HLA phenotypes. These cells provide a reference panel of antigens that are representative of the local population. Alternatively, panels may also employ preharvested or commercially provided frozen cell preparations, thereby avoiding the need for fresh lymphocyte isolations. In either case, testing is performed by a complement-dependent cytotoxicity assay.

CLL Panels
Frozen panels may utilize cells derived from individuals with chronic lymphocytic leukemia (CLL). These cells, being of the B lymphocyte lineage, provide a more sensitive indicator of antibody reactivity to both class I and class II HLA antigens. Autolymphocytotoxic antibodies, which mediate false positive reactions with the standard panels, are generally nonreactive with CLL panels.

ELISA Methods
Enzyme linked immunosorbent assays (ELISA) for the detection of HLA antibodies involve the use of purified HLA proteins as the target antigens. The detection system is usually specific for IgG class immunoglobulins and is not complement dependent.

SPECIMEN REQUIREMENTS

Serum derived from 10 ml of clotted blood (red top vacutainer tube) will suffice for all antibody testing methods. Typically, HLA antibodies are monitored monthly. Additional specimens should be obtained for testing two weeks following blood transfusions. When a transplant nephrectomy is performed as a result of graft rejection or failure, antibody testing should be performed at weekly intervals for four weeks.

INTERPRETATION

An elevated PRA is highly suggestive of the presence of HLA antibodies. The presence of HLA antibodies may be confirmed by CLL panel or ELISA reactivity. A positive PRA in the absence of CLL or ELISA reactivity suggests the presence of IgM class autoantibodies. Reactions can be attributed to IgM class immunoglobulins if they are neutralized by heat inactivation or chemical reduction. If the presence of a HLA antibody is confirmed and the antigenic specificity is determined, the corresponding antigen should be considered an unacceptable mismatch in all potential donors.

CROSSMATCH TESTING

APPLICATIONS

Crossmatch testing detects the presence of recipient antibodies that react specifically with the tissue of the allograft donor. These reactions serve as indicators of donor and recipient incompatibility and usually predict imminent antibody-mediated rejections.

INDICATIONS

Recipients of cadaveric renal and pancreas allografts must be crossmatched prospectively to transplantation. Heart, lung and liver recipients may be crossmatched retrospectively, provided that sensitization to HLA antigens has not occurred. Crossmatch testing for living related renal transplants should be performed prospectively in all cases. It is recommended that living related crossmatches be done twice and that final testing be performed 48 h to 7 days prior to surgery.

METHODS

Standard Cytotoxicity Method

The standard crossmatch procedure is a complement-dependent cytotoxicity assay that detects recipient antibodies to donor lymphocyte preparations.

Modified Crossmatch Methods

Variations of the standard crossmatch include the Amos cell washing modification, the antiglobulin enhancement procedure and extended incubation assays. These procedures improve the sensitivity of the basic cytotoxicity crossmatch. The use of enriched B cell preparations may also improve crossmatch sensitivity, particularly for the detection of antibodies to class II antigens.

Flow Cytometry

The flow cytometric crossmatch procedure detects donor reactive antibodies by indirect immunofluorescence. With the appropriate modifications, the assay is specific for IgG class antibodies that are directed against donor T cells or B cells.

SPECIMEN REQUIREMENTS

Recipient Specimens

Serum derived from 10 ml of clotted blood (red top vacutainer tube) is required of all potential allograft recipients. These specimens should be less than 30 days old at the time of transplant. If a potential sensitizing event, i.e., transfusion, has occurred within that period of time, a more recent specimen will be required. In patients with documented sensitization, the most reactive sera obtained within the previous 6 to 12 month period should also be tested. In regraft patients, it is recommended that serum collected at the time of rejection of the previous allograft be used for subsequent crossmatch testing.

Donor Specimens

For cadaver organ donors, it is most desirable that preharvest lymph nodes be used as the source of tissue cells. Crossmatch testing for living related transplants generally requires 30 ml of heparinized (green top vacutainer tubes) donor whole blood. The collection of donor serum (red top vacutainer tube) is also recommended to establish baseline reactivity in the flow cytometric crossmatch.

Interpretation

Absence of crossmatch reactivity predicts a favorable graft outcome. Reactivity in the standard or modified crossmatches, which is attributable to HLA antibodies, is a contraindication to transplantation. Reactions due solely to IgM class non-HLA antibodies are not considered clinically significant. A positive flow cytometric crossmatch should not be considered an absolute contraindication to transplantation, particularly in unsensitized patients. However, a significant risk of allograft rejection is associated with flow cytometric reactions in sensitized and regraft patients.

SELECTED READING

1   ASHI Clinical Affairs Committee, Guidelines for Clinical Histocompatibility Practice. ASHI Quarterly 1996; 15-17.
2   Braun, WE. Laboratory and Clinical Management of the Highly Sensitized Organ Transplant Recipient. Hum Immunol 1989; 26:245-260.
3   MacQueen JM, Barger BO. Histocompatibility Testing for Transplantation. *in* Organ Procurement, preservation and Distribution in Transplantation. Phillips MD, ed, William Byrd Press, Inc. 1991.
4   Mahoney RJ, Norman DJ, Colombe BW, Garovoy MR, Leeber DA. Identification of High and Low Risk Second Kidney Grafts. Transplantation 1996; 61(9):1349-1355.

5   Ting A, Welsh. HLS Matching and Crossmatching in Renal Transplantation. *In* Kidney Transplantation Principles and Practice. Morris PJ, ed, WB Saunders. 1994.

6   Zachary AA, Klingman L, Thorne N, Smerglia AR, Teresi GA. Variations of the Lymphocytotoxicity Test. Transplantation 1995; 60(5):498-503.

7

Renal Transplant Biopsy: The Pathologist's View

Edwin H. Jenis, Ihsan Housini

8

INTRODUCTION

The main indication for performing a kidney transplant biopsy is graft dysfunction, reflected by an increasing serum creatinine level. The most common causes of clinical graft dysfunction include allograft rejection, cyclosporine toxicity and recurrence of primary disease, which cannot be unequivocally distinguished by clinical criteria alone. The standard needle core biopsy is currently the definitive diagnostic test for the evaluation of allograft dysfunction, and for determining which of the many potential pathologic conditions is affecting the renal allograft. It is generally recommended that two needle core biopsies be obtained, since the false negative rate from a single core biopsy is approximately 10%. This is due to the fact that the rejection process may be patchy at its onset. The estimated false negative rate for two simultaneous needle core biopsies is less than 1%. A renal biopsy without cortical tissue, which consists only of renal medulla, is insufficient for assessing a renal allograft. The standard needle core biopsy is superior to fine needle aspirations, since in the latter technique blood vessels cannot be evaluated, the spatial relationships of the components of the kidney are lost, and the FNA technique cannot assess chronic allograft changes, de novo or recurrent glomerular diseases. This chapter, therefore, will only focus on the conventional needle core biopsy.

Kidney and Pancreas Transplantation, 2nd Edition, edited by Ingemar J.A. Davidson © 1999 Landes Bioscience

Table 8.1. *Example of clinical information to accompany a renal transplant biopsy used at the State University of New York at Buffalo Transplant Program*

Clinical Information: Renal Allograft Biopsy

    Patient Name: DOB: Age: Sex:

    Hospital: Date Obtained: Surg. Accession:

    Doctor Name: Phone/Beeper:

    Donor Name: Living Related: Cadaver: HLA Match:

Graft:

    1. Ischemia Time:

    2. Donor Biopsy: Y/N   Where: Date:

    3. Initial Function:

Recipient:

    1. Primary Renal Disease:

    2. Pretransplant Nephrectomy: Y/N

    3. First Transplant?   Second   Third

    4. Date of Current Transplantation:

    5. Previous rejection episodes this kidney?

                      <u>Date</u>   <u>Therapy</u>   <u>Response</u>

        a. First Episode

        b. Second Episode

        c. Third Episode

    6. Clinical Symptoms:

        a. Fever   b. Graft Tenderness   c. Hypertension

    7. Laboratory/Radiology:

        a. Serum Creatinine: Acute Rise   Chronic Rise   Failure to decline

        b. Proteinuria (g/24h)

        c. Active Urinary Sediment

        d. Urine Output

        e. Renal Scan

           Other Pertinent Findings:

    8. Clinical Impression:

        a. Acute Rejection   g. Renal Artery Stenosis

        b. Acute Tubular Necrosis   h. Severe Hypertension

        c. Chronic Rejection   i. Thrombotic Microangiopathy (HUS)

        d. Obstructive Uropathy   j. Recurrent GN

        e. Cyclosporine Toxicity   k. De Novo GN

        f. FK 506 Toxicity

           Other pertinent history/concerns:

    9. Current Medications/Immunosuppression with Dosages and/or Serum Levels:

        a. Azathioprine   d. ALG

        b. Prednisone   e. OKT 3

        c. Cyclosporine   f. FK506

           Other Meds (includes antibiotics)

   10. Has the patient been pulsed prior to biopsy? Y/N   Date:

HANDLING OF THE TRANSPLANT KIDNEY BIOPSY

Communication and coordination with the renal pathologist is the key component in obtaining the correct tissue diagnosis. Pertinent clinical history *must* be supplied (Table 8.1). The fixative of choice for allograft renal biopsies is Zamboni's fixative (paraformaldehyde and picric acid) (Stefani et al, 1967), which can be used for both light microscopy with special stains and electron microscopy. A portion of the renal biopsy should be placed in an immunofluorescent fixative (i.e. Michel's fixative). Light microscopy with special stains (i.e. H & E, PAS, and trichrome stains), and immunofluorescence microscopy for immunoglobulins G, A, M, C3, C4, fibrin and properdin are routinely performed. Immunohistochemical studies can be performed on the tissue submitted for light microscopy, and on quick frozen tissue for immunofluorescence microscopy, if required. The portion of tissue submitted for electron microscopy is especially useful if there is evidence of immune complex deposition in the portion of tissue submitted for immunofluorescence microscopy, or there is the possibility that the patient may be developing recurrent or de novo glomerulonephritis.

It is useful to obtain a biopsy of the donor kidney immediately after revascularization as a baseline for the evaluation of subsequent allograft biopsies.

FROZEN SECTION

Frozen sections of the donor kidney are technically limited to evaluation of the percentage of sclerosed glomeruli, the degree of interstitial fibrosis, the presence of significant vascular disease and the presence of malignancy. The findings by frozen section, therefore, are especially helpful in evaluating cadaver donor kidneys as part of the decision making process regarding their usefulness for transplantation. If the frozen section of the donor kidney reveals that greater than 20% of the glomeruli are sclerotic, the kidney is not suitable for transplantation.

Frozen sections of allografted kidneys can reveal the presence of interstitial inflammatory infiltrates and occasionally endothelialitis, cortical necrosis, or hyperacute rejection. Tubulitis and the presence of glomerular or vascular thrombi are difficult to identify on frozen sections, and most glomerular lesions are difficult to assess by this technique.

Renal biopsies for histological evaluation by light microscopy are usually reported within 24 h, but can be processed in 3 h if the appropriate tissue embedding systems and personnel are available. The differential diagnosis of rejection versus cyclosporine nephropathy can usually be rendered by light microscopy alone.

PATHOLOGY OF RENAL TRANSPLANTATION

The conditions most commonly identified in a renal transplant biopsy are indicated in Table 8.2. Clinically, the most common causes of renal dysfunction are allograft rejection or cyclosporine nephrotoxicity. Transplant rejection can be classified as hyperacute, accelerated, acute, or chronic. Acute and chronic changes may occur simultaneously.

Hyperacute transplant rejection is mediated by preformed circulating antibodies in the recipient against antigens present in the donor kidney. The circulating antibodies are generally directed against ABO and other blood group antigens,

Table 8.2. Diagnostic classification of renal allograft biopsies

1. Allograft Rejection:
   A. Antibody-mediated:
      1) Hyperacute allograft rejection
      2) Necrotizing arteritis (accelerated rejection)
      3) Thrombotic vasculopathy (HUS)
   B. T cell-mediated:
      1) Acute cellular allograft rejection
         i) Tubulointerstitial : 80%*
         ii) Vascular 50%*
         iii) Glomerular (Acute allograft glomerulopathy): 4%*
   C. Pathogenesis unknown or mixed humoral and cellular:
      1) Chronic allograft rejection
         i) Tubulonterstitial
         ii) Vascular
         iii) Glomerular (chronic allograft glomerulopathy)
2. Drug Toxicity:
   A. Cyclosporine nephrotoxicity
   B. FK 506 nephrotoxicity
   C. Acute allergic interstitial nephritis
3. Ischemic Injury:
   A. Acute tubular necrosis
   B. Perfusion injury
   C. Major vessel occlusion
4. Obstruction and Reflux
5. Infection:
   A. Bacterial, fungal, viral
6. Recurrent Primary Disease
7. De Novo Glomerulonephritis
8. Posttransplant Lymphoproliferative Disease

*% of cases of acute cellular rejection manifesting these changes

or against HLA antigens expressed on endothelial cells (see chapter 7). Hyperacute allograft rejection, an antibody-mediated process, can occur within minutes or hours after revascularization. The graft becomes edematous and discolored. With proper crossmatching techniques between the donor and the recipient, however, this type of rejection is rarely seen today (the editor has never experienced a hyperacute rejection during his 20 years experience as an active renal transplant surgeon).

Accelerated rejection (i.e. necrotizing arteritis) occurs 2-5 days after transplantation in less than 1% of allograft biopsies, is mediated by humoral antibodies and resembles microscopic polyarteritis nodosa. Approximately 75% of allografts expressing accelerated rejection are lost within one year.

ACUTE TRANSPLANT REJECTION

Acute transplant rejection is predominantly mediated ty T lymphocytes. All components of the kidney may be involved (i.e., tubules/interstitium, blood vessels and glomeruli). This process usually occurs within 6 weeks posttransplantation.

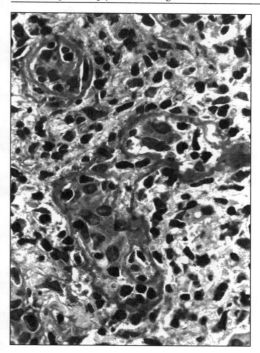

8

Fig. 8.1. Tubulitis is characterized by the presence of numerous mononuclear cells invading the walls of renal tubules.

Acute cellular *tubulo-interstitial* rejection is the most common pattern observed, and may be diffuse (80%) or focal (20%). Mononuclear cells first accumulate within the peritubular capillaries, then migrate into the interstitium, and finally invade the walls of the renal tubules, producing the lesion of tubulitis (Fig 8.1). In this location, the mononuclear cells and/or their cytokines produce tubular cell degeneration and necrosis, which ultimately results in tubular atrophy.

Acute cellular *vascular* rejection primarily involves small arteries and arterioles. The endothelial cells undergo swelling and vacuolization and are frequently sloughed from the intimal surface of the blood vessels. Mononuclear cells infiltrate the intima, which is referred to as endothelialitis (Fig 8.2). Endothelialitis is pathognomonic of acute cellular rejection, but is present in only 50% of kidneys undergoing allograft rejection. It is frequently a focal or patchy process.

Pathologic lesions that resemble acute cellular tubulo-interstitial rejection include herpes interstitial nephritis (characterized by typical nuclear inclusions), drug-induced interstitial nephritis (which does not manifest endothelialitis), other forms of viral interstitial nephritis and lymphomatous involvement of the allograft, which will not have an associated endothelialitis, and can be evaluated by appropriate immunohistologic techniques utilizing monoclonal antisera to lymphocyte markers.

8

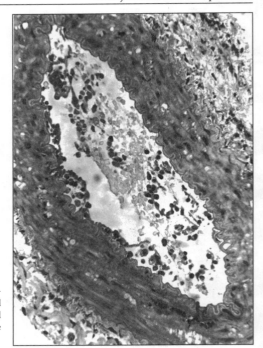

Fig. 8.2. Endothelialitis is characterized by the presence of endothelial cell degeneration, endothelial cell sloughing, and a subintimal infiltrate of mononuclear cells.

Acute cellular *glomerular* rejection (i.e. acute allograft glomerulopathy) occurs in approximately 5% of cases of allograft rejection and is characterized by endothelial cell swelling, capillary luminal obliteration and an accumulation of T cells within the glomerular tufts.

CHRONIC TRANSPLANT REJECTION

Chronic allograft rejection, characterized clinically by a slowly increasing serum creatinine level, increasing proteinuria and hypertension, is characterized by greater or lesser degrees of tubular atrophy, interstitial fibrosis, a variable interstitial inflammatory infiltrate, vascular lesions characterized by varying degrees of intimal fibroelastosis, and vascular medial fibrosis. Changes of acute cellular rejection may be superimposed on, or occur concomitant to the changes of chronic allograft rejection. Chronic allograft glomerulopathy (i.e., transplant glomerulopathy) is characterized by diffuse or focally sclerotic glomeruli with thickened capillary walls, demonstrating double contoured lesions on special stains.

CYCLOSPORINE NEPHROTOXICITY

Cyclosporine induces preglomerular arteriolar vasoconstriction which is not associated with pathologic changes in the early stages of the process. The earliest

Fig. 8.3. Acute cyclosporine toxicity characterized by isometric vacuolization of the cytoplasm of renal tubules.

8

pathologic changes which characterize acute cyclosporine nephropathy include isometric vacuolization of the cytoplasm of the renal tubules (Fig. 8.3) and focal areas of dystrophic microcalcification within the tubular lumina.

Acute cyclosporine vascular injury is characterized by acute focal myocyte lesions (i.e., individual smooth muscle cell degeneration, vacuolization and necrosis in the media of small arteries). Acute cyclosporine nephropathy can also produce the clinical and pathologic manifestations of the hemolytic uremic syndrome. Pathologically, thrombi are identified within small arteries, arterioles and glomeruli with accompanying extravasated red blood cells and schistocytes. FK 506 can also produce the hemolytic uremic syndrome with pathologic changes identical to those produced by cyclosporine. The pathologic changes of the hemolytic uremic syndrome produced by cyclosporine or FK 506 are not distinguishable from allograft rejection associated hemolytic uremic syndrome, recurrent hemolytic uremic syndrome, or thrombotic thrombocytopenic purpura. The only unequivocal evidence of rejection is endothelialitis, which is not found in cyclosporine or FK 506 toxicity alone. Since cyclosporine toxicity correlates imperfectly with blood or serum cyclosporine levels, the allograft renal biopsy is currently the most sensitive technique for resolving the differential diagnosis of acute cellular rejection from acute cyclosporine nephropathy.

8

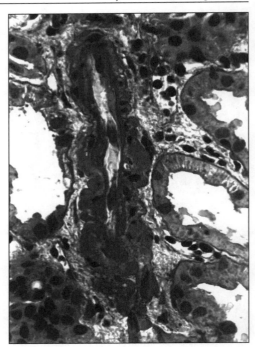

Fig. 8.4. Chronic cyclosporine toxicity characterized by arteriolar
medial hyalinosis.

Chronic cyclosporine nephropathy is characterized by arterial and arteriolar
medial hyalinosis (Fig 8.4), arterial mucoid intimal thickening, tubular atrophy
and "striped" interstitial fibrosis. The medial hyalinosis seen with cyclosporine
and FK 506 toxicity may be progressive, or may regress in approximately 50% of
cases if the dosage of cyclosporine or FK 506 is decreased, or treatment is totally
discontinued.

OTHER PATHOLOGIC CHANGES AFFECTING
THE TRANSPLANTED KIDNEY

Acute tubular necrosis (ATN) is a common cause of delayed primary graft
dysfunction due to ischemia of the graft during harvesting, preservation, or transplantation. Histologically, ATN is characterized by sloughing of tubular cells with
necrotic debris in the tubular lumina. Regenerative epithelial changes with mito-
ses are seen in later stages.

The diagnosis of recurrence of the original kidney disease within the allograft
can be made only if the previous diagnosis has been documented thoroughly by

light microscopy, immunofluorescence microscopy and electron microscopy in an early stage of its development. Recurrent glomerulonephritis occurs in approximately 1-5% of allografts. The forms of glomerulonephritis that most commonly recur (in decreasing order) are: basement membrane dense deposit disease, the hemolytic uremic syndrome, IgA nephropathy, focal segmental glomerulosclerosis and the nephritis associated with the Henoch-Schönlein syndrome. Although most of these recurrent diseases have relatively little effect upon the survival of the allograft, recurrent focal segmental glomerulosclerosis is commonly associated with graft loss. Focal segmental glomerulosclerosis can recur as early as a few hours to days after transplantation and may result in rapid loss of the transplanted kidney.

De novo glomerulonephritis occurring within the allografted kidney, is relatively uncommon. The most common form of glomerulonephritis occurring de novo is membranous nephropathy, which does not significantly shorten graft survival. Anti-GBM-mediated glomerulonephritis most commonly develops in patients with Alport's syndrome receiving a normal donor kidney. Although the Alport recipients frequently develop linear immunoglobulin G deposits on the glomerular capillary basement membranes, most of these patients do not develop overt glomerular injury.

8

PERFUSION NEPHROPATHY

Perfusion nephropathy resembles hyperacute allograft rejection pathologically, but is not associated with preformed circulating antibodies. Perfusion nephropathy results from endothelial damage during pulsatile perfusion of the graft before implantation, or occurs in kidneys exposed to prolonged cold ischemia time. The glomeruli are typically infiltrated by neutrophils, and there is associated endothelial cell swelling and degeneration with associated glomerular thrombi. These morphologic changes cannot be differentiated from hyperacute rejection, and therefore the differential diagnosis must be resolved by clinicopathologic correlation.

POSTTRANSPLANT LYMPHOPROLIFERATIVE DISEASE

Posttransplantation lymphoproliferative disease is characterized by a diffuse mononuclear cell infiltrate within the renal transplant. The pathologic lesions range from benign interstitial lymphoid proliferations with mononucleosis type symptoms to monoclonal proliferations resembling and behaving like a non-Hodgkin's lymphoma. The latter are usually of B lymphocyte origin and may involve extranodal sites. Most are associated with evidence of Epstein-Barr virus infection.

SELECTED READING

1   Colvin RB. The renal allograft biopsy. Kidney Int 1996; 50:1069-1082.
2   Kolbeck PC. Renal transplant pathology. In: Kolbeck PC, Markin RS, McManus BM, eds. Transplant Pathology. Chicago: ASCP Press, 1994:161-195.
3   Meehan SM, Colvin RB. Differential diagnosis of renal allograft biopsies. In: Hammond EH, ed. Solid Organ Transplantation Pathology. Philadelphia: WB Saunders Co, 1994:159-185.
4   Stefani M, DeMartino C, Zamboni L. Fixation of ejaculated spermatozoa for electron microscopy. Nature 1967; 216:173-174.
5   Striker G, Striker LJ, D'Agati V. Transplantation. In: The Renal Biopsy. 3rd ed. Philadelphia: WB Saunders Co, 1997:242-257.

Percutaneous Interventions in Kidney and Kidney/Pancreas Transplants

George L. Miller, Michael J. Wallace

ANGIOGRAPHY

Technologic advancements in diagnostic radiology over the past 10-15 years have decreased the role of angiographic evaluation of renal and pancreatic transplants. Preoperative donor renal arteriography is currently the modality of choice for identifying kidney size as well as the number, length and location of renal arteries and veins prior to transplantation. Anomalous or accessory renal arteries may arise from any location extending from the supraceliac aorta to the iliac arteries (Fig. 9.1). As a byproduct of angiography, contrast within the collecting system provides pelvic caliceal anatomy. Additionally, renal angiography is necessary for the identification of intrinsic renal arterial disease such as fibromuscular dysplasia and atherosclerotic stenosis. While angiography remains the gold standard for preoperative donor evaluation, CT and MR angiography may, in the future, provide a noninvasive means for obtaining the needed information.

Radionuclide imaging and ultrasound studies have now supplanted arteriography as primary diagnostic tools in the differentiation of rejection, acute tubular necrosis and obstruction. Angiography is now used primarily for the evaluation of potential arterial complications including arterial stenoses, venous stenoses, occlusions, and pseudoaneurysms, which may contribute to graft dysfunction or hypertension. Arterial stenoses are the most common vascular complication, occurring in 4-12% of transplant renal arteries. With continued improvement of Doppler ultrasound, magnetic resonance angiography and spiral CT angiography and their ability to display arterial and venous anatomy, the need for angiography may soon be limited to those cases that require therapeutic intervention (i.e. balloon angioplasty). These noninvasive diagnostic modalities can be advantageous in patients with graft dysfunction and markedly elevated creatinine levels when

9

Kidney and Pancreas Transplantation, 2nd Edition, edited by Ingemar J.A. Davidson © 1999 Landes Bioscience

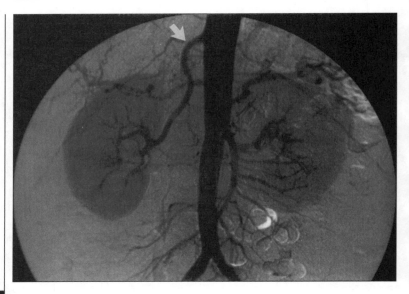

Fig. 9.1. Digital subtraction arteriogram with an anomalous right renal artery arising from the supraceliac aorta (arrow).

9

the use of iodinated contrast material becomes an issue. Alternative contrast agents like carbon dioxide and gadolinium have recently been employed in selected situations with some success.

Before undertaking transplant arteriography, it is useful to have an understanding of the variety of surgical anastomoses that are commonly utilized. The surgical techniques and common sites of transplant artery anastomoses are discussed in chapter 2, Figure 2.15. Two common anastomoses encountered in our practice are the Carrel patch to the external iliac artery and the end-to-end anastomosis with the internal iliac artery. Regardless of the anastomosis which has been utilized, the approach for diagnostic arteriography is usually from the contralateral femoral artery. If the femoral arterial pulse is diminished or absent, the ipsilateral approach should be undertaken. In the latter circumstance, care must be taken with initial access to prevent arterial spasm or intimal dissection at the transplant anastomosis. Injury is avoided by fluoroscopically monitoring the passage of the guidewire through the iliac arterial system to insure that it does not inadvertently enter the transplant artery. If spasm is encountered, intra-arterial nitroglycerin or oral/sublingual nifedipine can be useful.

RENAL ARTERY STENOSIS
In patients with renal transplants and possible renovascular hypertension, initial attention is directed toward the distal lumbar aorta and proximal pelvic arter-

ies. This maneuver is to exclude the presence of any hemodynamic stenoses involving inflow that may result in transplant hypoperfusion and renovascular hypertension. Following inflow evaluation, the anastomosis with the native artery is evaluated in multiple projections in order to display the region in profile. Since the course of the transplant renal arteries are usually foreshortened in the AP projection, steep oblique views are often needed for optimum visualization (Fig. 9.2).

If a stenosis of either the native or transplant arteries is discovered in a patient with severe hypertension of recent onset, dilation of the lesion can be undertaken with standard percutaneous techniques. This may include angioplasty alone or in combination with stenting. Depending on the type of anastomosis and its orientation with the native artery, a second femoral arterial access contralateral to the initial access site may be necessary to provide the most advantageous route to the lesion. Resistant stenoses, particularly of the transplant artery anastomosis may require high pressure balloons to achieve successful dilation. Angioplasty has

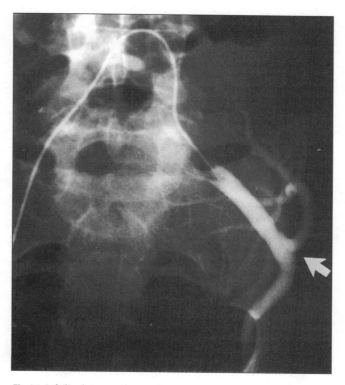

9

Fig. 9.2. Left iliac fossa transplant renal arteriogram from the right common femoral approach. The angiographic catheter is positioned within the contralateral (left) iliac artery. Note the widely patent anastomosis (arrow) displayed in profile.

become the first line of therapy for transplant or inflow stenoses identified in patients with suspected renovascular hypertension. Vascular stents should be reserved for salvage situations following suboptimal angioplasty. Complications associated with percutaneous transplant artery angioplasty are uncommon, but may include occlusive intimal dissection, in situ thrombosis vessel rupture, or distal embolization of debris from the treatment site.

ARTERIO-VENOUS FISTULA

An additional arterial complication that arises with relative frequency is the arterio-venous fistula (Fig. 9.3). This particular abnormality usually arises as a

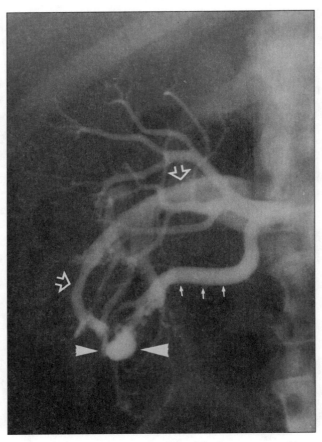

Fig. 9.3. Right renal arteriogram in a native kidney following a percutaneous biopsy. Note the sizable arterio-venous fistula with the supplying artery (small arrows), a pseudoaneurysm component (arrow heads) and early venous opacification (open arrows).

result of percutaneous biopsies of the transplant kidney as part of the work up for rejection. These lesions may also be associated with a pseudoaneurysm component. Patients may present with hematuria, a peritransplant hematoma and/or a hemoglobin/hematocrit decline below baseline. With the dramatic advancement in catheter materials and the development of microcatheters, it is now possible to advance these devices peripheral within the renal parenchyma and embolize only a small territory of parenchyma. The goal in this situation is to preserve as much functioning tissue as possible.

PERCUTANEOUS BALLOON DILATION
OF COLLECTING SYSTEM OBSTRUCTIONS

Complications involving the renal collecting system may also be a contributing factor in transplant dysfunction. Obstruction can occur early due to necrosis or can be technical in nature involving the ureteroneocystostomy implantation site (chapter 11, Fig. 11.8). Late causes of obstruction may be secondary to a ureteral stricture, stone or tumor formation. Usually, initial maneuvers to bypass the obstruction include the retrograde insertion of a ureteral stent under cystoscopic guidance. In approximately half of the cases, the latter approach is unsuccessful ($\approx$50%) or contraindicated. External drainage via a percutaneous nephrostomy tube can then provide initial decompression. Insertion of a transplant nephrostomy tube into an obstructed and dilated collecting system is a relatively straightforward procedure. Given the variable location and orientation of renal transplants as well as its superficial location, the approach must be tailored to the individual patient.

The basic technique for transplant nephrostomy tube insertion is similar to that used for native kidneys, aside from the patient being placed in the supine position. In situations where graft dysfunction prohibits the use of intravenous contrast, ultrasound or CT guidance can be used to identify and access the collecting system (chapter 11, pg. 195). If contrast can be used, a simple intravenous injection of 50-100 cc will opacify the collecting enough to allow access under fluoroscopic guidance. Once the access needle has entered the chosen infundibulum or calyx, a guidewire is passed into the renal pelvis and/or proximal ureter. Fascial dilators are then used to dilate the tract to 8 French and then exchanged for a retention type of nephrostomy catheter with its loop formed and locked within the renal pelvis. These catheters are left to gravity drainage and the urinary output monitored subsequently.

Antegrade ureteral stents can be placed in transplant kidneys by a similar technique to that utilized in the native collecting system. The major difference is the orientation between the transplant kidney and native bladder and the shortened ureteric length. The latter difference is overcome by using special shorter ureteral stents.

THE EVALUATION OF PANCREAS TRANSPLANTS

Like renal transplants, the pancreas has an excretory function and requires some type of output reservoir. Exocrine duct diversion in pancreatic transplants are usually to the recipient bladder or a loop of small bowel. The surgical consideration for this procedure is discussed in chapters 1 and 3. In those patients in whom the bladder has been utilized as a reservoir, the radiologic evaluation of the anastomotic complications can be evaluated by retrograde cystography.

Following scout views of the pelvis in various obliquities, the bladder is catheterized and filled by gravity with dilute water soluble contrast material with intermittent fluoroscopic monitoring. When the patient experiences a sensation of fullness in the bladder, the flow of contrast is stopped and films are obtained in multiple projections to display the anastomosis in profile (Fig. 9.4). Spot films are used to augment the examination so that the anastomotic integrity and patency may be adequately evaluated. The duodenocystostomy should fill easily (Fig. 9.5) and rarely contrast may be seen refluxing into the pancreatic duct. The demonstration of contrast extravasation from the anastomosis or the duodenal stump (Fig. 9.6) is abnormal and indicates the presence of an anastomotic breakdown and leak, usually associated with significant clinical symptoms (see also chapter 3, Fig. 3.13).

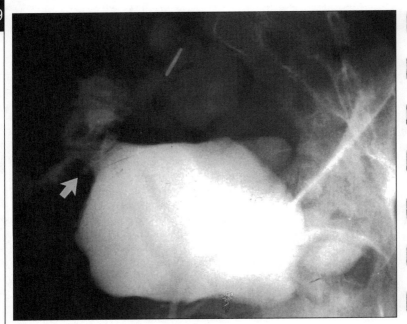

Fig. 9.4. Contrast cystogram in a steep oblique projection demonstrating the duodenocystostomy in profile. Note the anastomotic suture line (arrow). See also chapter 3, Figure 3.13c.

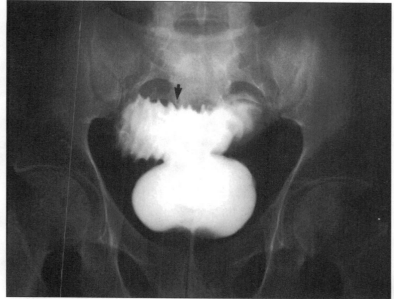

Fig. 9.5. AP cystogram with normal reflux into the duodenocystostomy. Note the well-opacified transplant duodenal segment with normal mucosal fold pattern (arrow). See also chapter 3, Figure 3.13c.

9

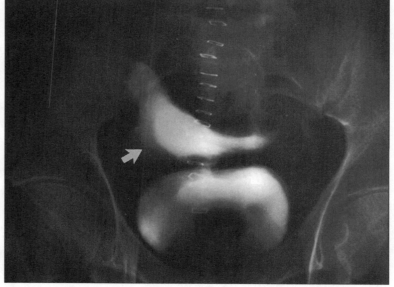

Fig. 9.6. AP cystogram demonstrating contrast extravasation (arrow) consistent with a leak. See also chapter 3, Figure 3.13a.

Angiographic evaluation of pancreas transplants is not routine, but rarely we have been asked to perform catheter angiography to evaluate for pancreatic transplant artery patency (Fig. 9.7). Communication with the transplant team as to the surgical anatomy is absolutely necessary so that the appropriate approach can be undertaken and the desired information is obtained.

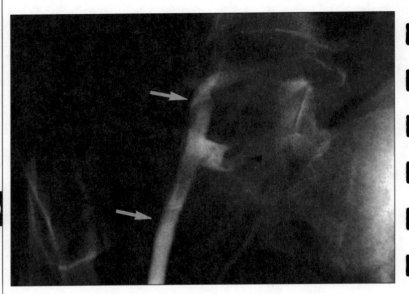

Fig. 9.7. Pelvic arteriogram with contrast opacifying the right iliac artery (arrows) and a stump of an occluded pancreatic transplant artery (arrow heads).

Ultrasonography of the Kidney and Pancreas Transplant

Anthony Setiawan

INTRODUCTION

Ultrasound is a practical and effective modality to evaluate patients with renal and pancreatic transplants. It is a noninvasive, quick and relatively inexpensive test. No radiation or potentially nephrotoxic contrast media are required. If needed, the examination can be done at the bedside.

The transplanted organs are relatively close to the anterior abdominal wall. There is little overlying bowel gas and therefore, they are easily seen in their entirety. For the same reason, high frequency transducers with excellent resolution and Doppler sensitivity can be used.

10

TERMINOLOGY

Echo: A reflection of a sound wave at an interface of two medias/tissues with different acoustic properties creates an echo. Each dot on an ultrasound image represents a single echo.

Echogenic: A structure filled with many echoes is echogenic. The echogenicity of a structure correlates to the heterogeneity of its tissue. Solid organ/masses are generally echogenic.

Anechoic: A structure containing no echoes in it is anechoic. Simple fluids (e.g. transudate, urine, bile, etc.) are examples of anechoic materials.

Hypoechoic: A structure less echogenic than another is hypoechoic. The normal kidney, for instance, is hypoechoic in comparison to the liver.

Isoechoic: Two structures that have the same degree of echogenicity are isoechoic.

Kidney and Pancreas Transplantation, 2nd Edition, edited by Ingemar J.A. Davidson © 1999 Landes Bioscience

Acoustical Shadow: An artefact due to total reflection at a strong interface (e.g. calcification). An anechoic area is seen distal to the interface (i.e. acoustical shadow).

Reverberation Artefact: An artefact that is seen distal to a very strong reflector (e.g. gas and metal), Sound waves that are reflected back and forth between the reflector and the transducer create this artefact. This is also known as ringdown artefact.

Doppler Effect: The frequency shift of a sound wave that is reflected from a moving object. The frequency increases when the object approaches the transducer and decreases when it moves away from it. The frequency shift can be displayed in a graphic form (spectral wave) or in color Doppler.

Resistive Index: The ratio (usually calculated electronically) of the peak systolic velocity minus the end diastolic velocity and divided by peak systolic velocity. This index reflects the peripheral vascular bed impedance.

Acceleration Time (AT): The time interval between the start of early systole to its peak. A prolonged AT (> 0.07 seconds) indicates *pulsus tardus.*

Acceleration Index (AI): A ratio that evaluates the slope of the early systolic peak.
- AI = Δ V/AT (See Fig. 10.1)
- A decrease in AI (< 300 cm/sec^2) indicates *pulsus parvus.*

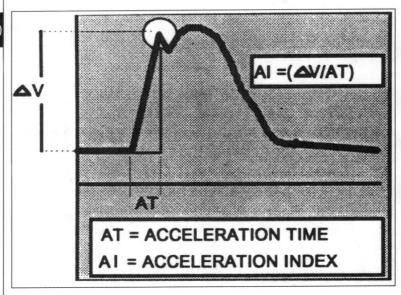

Fig. 10.1. Acceleration Index (AI): A ratio that evaluates the slope of the early systolic peak.

THE RENAL TRANSPLANT

NORMAL ANATOMY

In longitudinal sections (Fig. 10.2a) the transplanted kidney is seen as a bean shaped structure measuring 9 to 13 cm long (smaller if the donor is a child). An elongated, irregular echogenic structure caused by the fat around the renal pelvis and calyces (renal sinus), is seen in the center. On transverse images (Fig. 10.2b), this creates a doughnut or target appearance. The renal pelvis and calyces are usually completely collapsed and not generally seen. The renal pyramids (medulla) are seen as ill-defined triangular or round hypoechoic structures surrounding the renal sinus. The renal cortex consists of a rim of tissue, hyperechoic compared to the medulla, but less echogenic than the renal sinus.

The extrarenal vessels (Fig. 10.3) are usually seen posterior and lateral to the transplanted kidney and are often quite tortuous. The intrarenal branches in the renal sinus must not be confused with mild hydronephrosis (Fig. 10.4). Interlobar arteries can be seen as echogenic lines demarcating the pyramids. Arcuate arteries are sometimes seen at the corticomedullary junctions (Fig. 10.2a,2c). The intralobar vessels can only be seen using sensitive color Doppler (Fig. 10.2c).

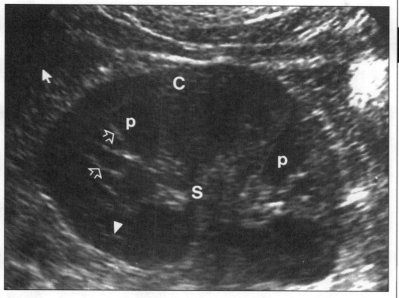

10

Fig. 10.2. Normal Anatomy. a) Longitudinal section of a kidney transplant demonstrates the renal medulla (pyramids) (P) that is hypoechoic in comparison to the renal cortex (C). The most echogenic central structure represents the renal sinus fat (S). Interlobar (open arrows) and arcuate vessels (arrow heads) are occasionally seen.

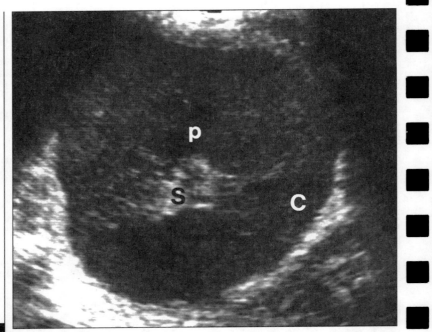

Fig. 10.2b. Transverse section demonstrates the characteristic doughnut or target appearance of the kidney.

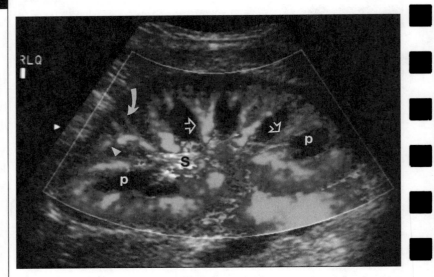

Fig. 10.2c. Power color Doppler image of the kidney transplant. Using this sensitive technique, interlobar (open arrows), arcuate (arrow head) and intralobular (curved arrow) vessels are easily seen.

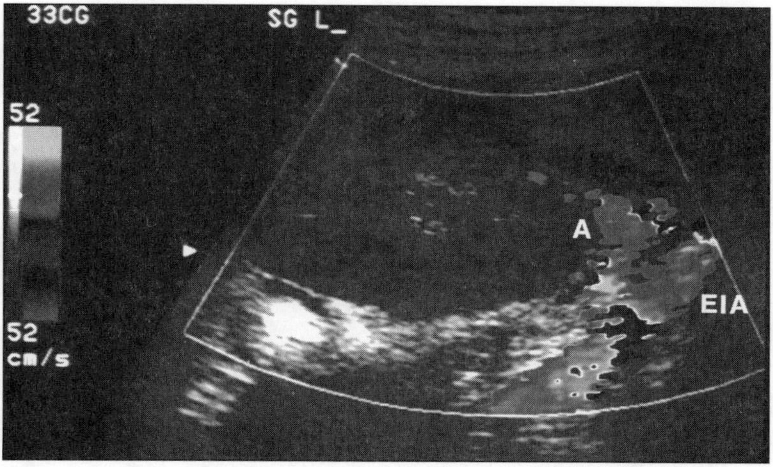

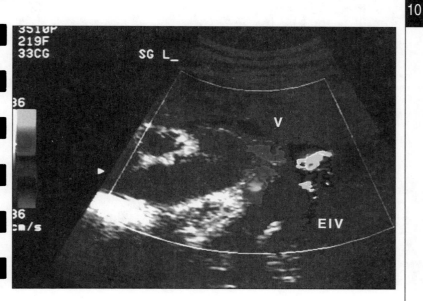

Fig. 10.3. Main extrarenal vessels imaged with color Doppler. a) Renal artery (A). EIA = External iliac artery.

Fig. 10.3b. Renal vein (v); EIV = External iliac vein.

10

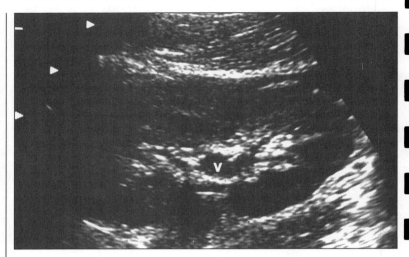

Fig. 10.4a. A prominent intrarenal vein (V) can be mistaken for mild hydronephrosis on gray scale imaging.

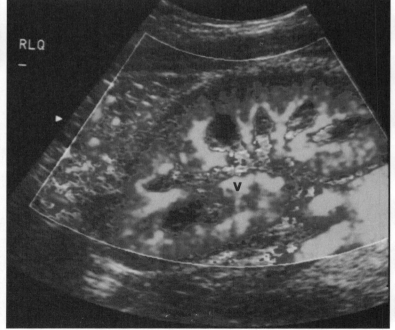

Fig. 10.4b. The correct diagnosis was made using Doppler evaluation.

HYDRONEPHROSIS

The most common ultrasound study requested for a renal transplant patient with an elevated serum creatinine is to rule out hydronephrosis indicative of an obstruction. Early dilation is seen as an anechoic tubular structure in the renal sinus which represents the intrarenal pelvis (Fig. 10.5a). In more advanced hydronephrosis, dilated calyces are seen branching off from the renal pelvis (Fig. 10.5b,5c). A dilated ureter can also be seen.

Pitfalls

Physiologic

A small transient urine collection can be seen in the intrarenal pelvis without extension into the calyces. This should not be confused with minimal hydronephrosis.

Postobstructive Dilatation

After the cause of obstruction has been resolved, the collecting system may remain dilated for some time. These changes can be permanent following a prolonged obstruction.

Reflux

Using ultrasound images alone, dilatation of the renal pelvis and calyces due to vesicoureteral reflux cannot be differentiated from dilatation due to obstruction. The presence or absence of ureteral jets of urine flow into the bladder, which can be seen with color Doppler, can differentiate the two conditions.

Full Bladder

Mild dilatation of the renal pelvis may be present in a patient with a full urinary bladder and should resolve on a postvoid examination (Fig. 10.6). The distended urinary bladder is easily recognized on the ultrasound examination.

Diuresis

Following a forced diuresis, (e.g., after IV furosemide) mild to moderate transient hydronephrosis can be seen, resolving within 24 h.

Acute obstruction

In early urinary tract obstruction (less than 12 h) there may not be enough pressure in the collecting system to cause distension that can be recognized with ultrasound.

Hydronephrosis Filled with Echogenic Material

Hydronephrosis can be hard to recognize due to substances other than urine (e.g. blood clot, large stones, etc.) that can fill a dilated renal pelvis (Fig. 10.7).

Despite the high sensitivity (98%) of ultrasound in detecting the presence of hydronephrosis (Ellenbogen et al, 1978) it is important to remember that a dilated collecting system is not equivalent to an obstruction. The resistive index of the intrarenal arteries may have some value in differentiating an obstructed from a nonobstructed but dilated collecting system (Platt et al, 1989). While the value is still controversial (Chrysi et al, 1990), a resistive index of less than 0.70 suggests nonobstructive dilatation, while values 0.70 and higher are suspicious for obstruction. An increasing amount of dilatation on serial ultrasound examinations is generally diagnostic of obstruction.

10

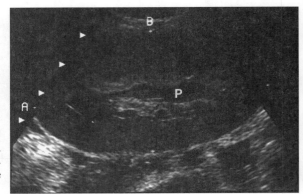

Fig. 10.5. Grades of hy-
dronephrosis. a) Mini-
mal dilatation of the
intrarenal pelvis (P).

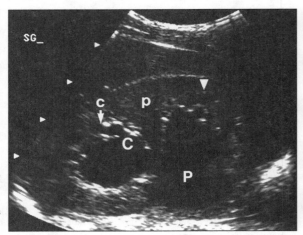

10

Fig. 10.5b. Moderate di-
latation with a branch-
ing pattern due to dis-
tention of major calyces
(C).

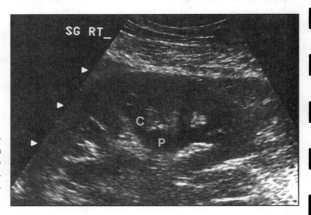

Fig. 10.5c. Severe dilatation with compression
of the renal cortex. c =
minor calyx; p = pyra-
mid; arrowhead = arcu-
ate artery

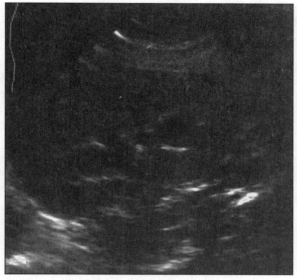

Fig. 10.6. False positive hydronephrosis (nonobstructive dilatation) due to a distended urinary bladder. a) Moderate hydronephrosis.

10

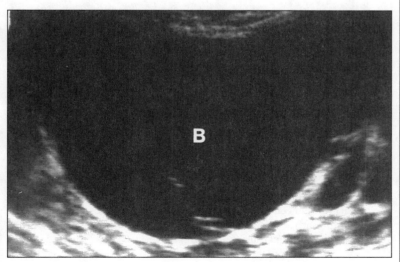

Fig. 10.6b. Distended urinary bladder (B).

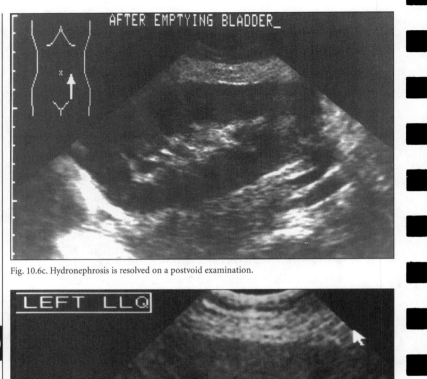

Fig. 10.6c. Hydronephrosis is resolved on a postvoid examination.

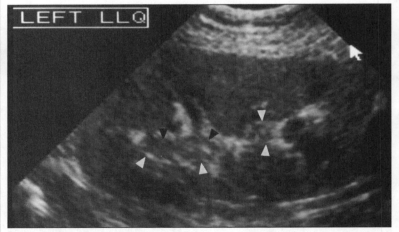

Fig. 10.7. Sonogram of a patient with gross hematuria following renal transplant biopsy. Hydronephrosis is somewhat difficult to see due to an echogenic blood clot (arrowheads) that fills the dilated renal pelvis.

FLUID COLLECTIONS

A loculated fluid collection is seen as a rounded anechoic mass. This differs from free fluid (such as ascites or fresh blood) that insinuates itself between bowel loops and organs without causing any real mass effect. Since a distended urinary bladder has the same appearance as a loculated fluid collection, it is essential that a postvoid repeat ultrasound is performed when such a collection is seen (Fig. 10.8).

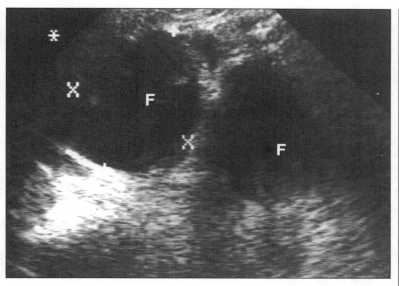

Fig. 10.8a. A left sagittal scan of the pelvis demonstrates two fluid collections next to each other (F).

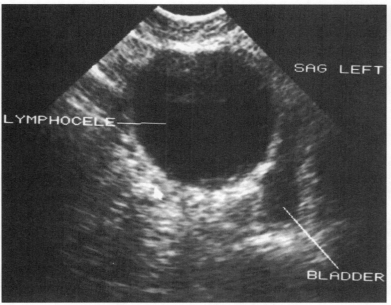

Fig. 10.8b. A postvoid study indicates which of these collection is abnormal.

10

In general, the nature of a fluid collection cannot be determined with ultrasound, but certain findings may suggest the diagnosis. A hematoma is usually found close to the transplanted kidney following a percutaneous biopsy or surgery. Acute hematomas are anechoic and, therefore, easily recognized. An older or subacute hematoma contains echoes and can appear isoechoic with the kidney making it harder to detect (Fig. 10.9a). Older liquefied blood collections are anechoic with linear echoes or dependent debris (Fig. 10.9b). Lymphoceles are the most common cause of peritransplant fluid collections and usually develop 1-3 weeks after transplant surgery. Lymphoceles are usually anechoic, but in chronic collections, linear echoes and dependent debris (due to thick proteinaceous material) may be present (Fig. 10.10). Urinomas often develop close to the ureterovesical junction relatively remote from the kidney. It is, therefore, important that the suprapubic area is routinely studied during ultrasound evaluation of the transplant. Any fluid collection can become infected without demonstrating any characteristic ultrasound findings. If gas collections develop, strong focal echoes with reverberation artefacts can be seen (Fig. 10.11). Since ultrasound findings are usually not specific, definite diagnosis of a fluid collection requires percutaneous aspiration (see "Aspiration–drainage").

REJECTION

Despite the major advances in image quality and new abilities of ultrasound, the diagnosis of rejection using morphologic changes remains unsatisfactory. For the past ten years, Doppler ultrasound has been evaluated extensively for diagnosing rejection and differentiating it from other causes of poor renal function of the kidney transplant. The decrease of diastolic flow in the interlobar and arcuate arteries due to increased impedance of the allograft microcirculation is the basis of this method (Rigsby et al, 1986 and Rifkin et al, 1987). In the most severe cases, absent (Fig. 10.12) or reversed (Fig. 10.13) diastolic flow can be seen. The degree of the reduced diastolic flow is measured with the resistive index. At our institution, we investigate only the interlobar arteries. The vessels in the upper, mid-and lower pole are imaged, using color Doppler and the resistive indexes are electronically calculated. A resistive index of 0.7 or less is considered normal. Resistive index values of 0.90 and higher are strongly suggestive of acute rejection. Resistive index values between 0.7 and 0.9 are indeterminate and can be due to causes other than rejection (acute tubular necrosis, cyclosporine toxicity, acute pyelonephritis and obstruction).

There is a significant difference of opinion as to the value of Doppler ultrasound in diagnosing acute rejection (Grant et al, 1990 and Genkins et al, 1989). Absence or even reversal of diastolic flow has been seen in severe cases of acute tubular necrosis, cyclosporine A toxicity and renal vein thrombosis (Kelcz et al).

Although not very specific, these spectral waveform abnormalities indicate a significant parenchymal problem (Middleton et al, 1989). Serial resistive indexes appear to have better specificity in diagnosing acute rejection compared to a single reading. A decrease of resistive index over time suggests clinical improvement. On

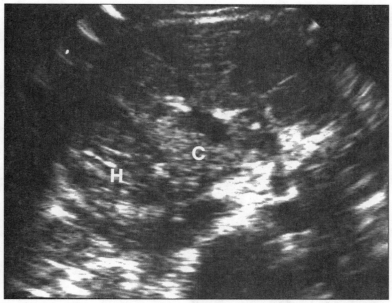

Fig. 10.9. Perirenal hematoma. a) In the subacute phase, a clotted hematoma (H) is echogenic, similar to the renal cortex (C). This may cause difficulties in detection.

10

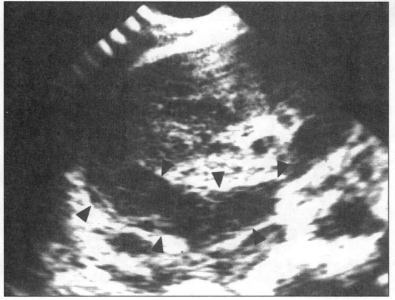

Fig. 10.9b. A chronic liquefied hematoma (arrowheads) is easily recognized due to its anechoic appearance.

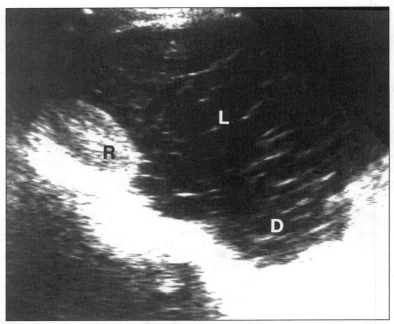

Fig. 10.10. Chronic lymphocele. A large lymphocele (L) with dependent debris (D) is seen medial to the renal transplant.

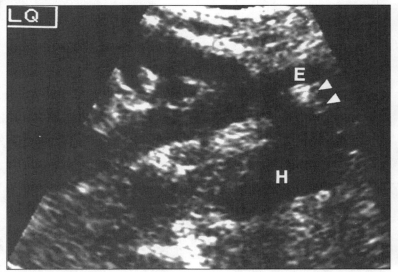

Fig. 10.11. Infected hematoma. A dense focal echo (E) with distal reverberation artefact (arrow heads) in a postop hematoma (H) indicates a gas collection. Infection due to gas forming organism was diagnosed with aspiration.

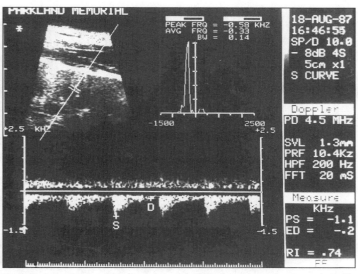

Fig. 10.12. Acute rejection. a) Baseline resistive index (RI) of a renal transplant in a patient with normal renal function.

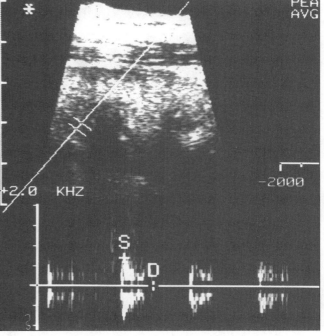

Fig. 10.12b. Absence of diastolic flow during an episode of severe rejection.

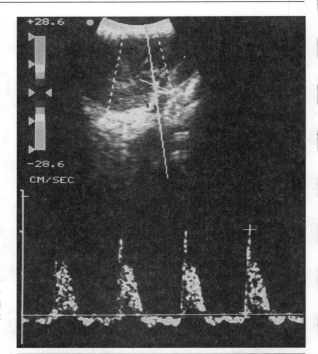

Fig. 10.13. Severe rejection. a) Reverse diastolic flow in a renal transplant in a patient with severe rejection.

10

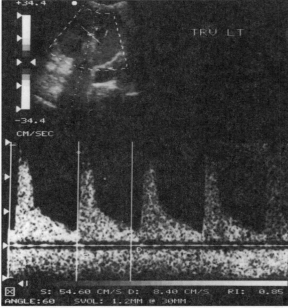

Fig. 10.13b. Following anti-rejection treatment, diastolic flow returns to an antegrade direction.

rare occasions, severe acute rejection with submucosal edema and hemorrhage can cause hypoechoic thickening of the renal pelvic wall (Fig. 10.14).

Cortical perfusion studies using power color Doppler and ultrasound contrast agents are a few of the new modalities that are currently under investigation for diagnosis of rejection.

INFECTION

Usually, no sonographic changes are seen in acute pyelonephritis. Severe edema can cause enlargement of the kidney and moderate increase of the resistive index. Focal nephritis (phlegmon) can cause an ill-defined hypoechoic "mass" (Fig. 10.15) while a more advanced renal abscess appears cystic with a "shaggy" wall. Severe edema of the renal pelvis can occasionally be seen as a hypoechoic thickening. This cannot be distinguished from the changes that can be seen in acute rejection. Fungus balls caused by *Candida albicans* can be seen as echogenic nodules in the renal pelvis (Fig. 10.16). In pyonephrosis, echoes can be seen within the fluid distending the collecting system. These echoes are indistinguishable from blood clots, but the clinical presentation usually indicates the diagnosis.

Vascular Complications

Renal Artery Stenosis

Direct and indirect methods of evaluating the renal artery for stenosis have been used.

The Direct Method

This requires evaluation for an increased peak systolic velocity at, and immediately distal to, the stenosis and for poststenotic turbulence (Fig. 10.17a, 17b).

10

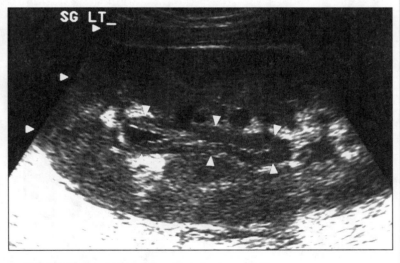

Fig. 10.14. Thickened renal pelvis wall in acute rejection (arrow heads).

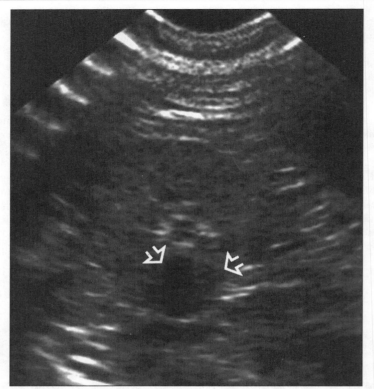

Fig. 10.15. Focal nephritis (open arrows). This differs from an anechoic and better defined (liquefied) abscess.

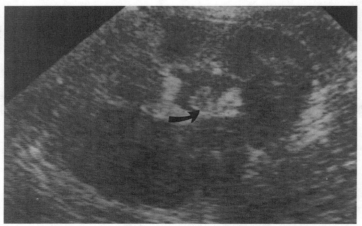

Fig. 10.16. Fungal infection. An echogenic filling defect in the renal pelvis (curved arrow) most likely represents a large fungus ball in this patient with *Candidiasis* of the urinary tract.

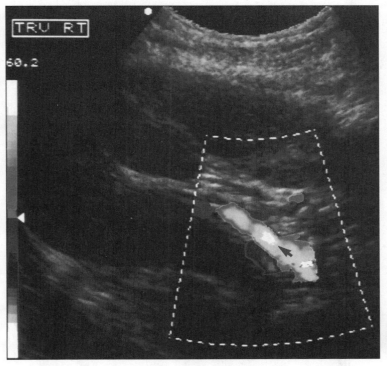

Fig. 10. 17. Renal artery stenosis diagnosed using the direct method. a) The white "spot" in the main renal artery (short arrow) indicates a focal area of increased flow.

10

These are the two main diagnostic criteria of renal artery stenosis. A perivascular tissue vibration can cause an unusual artefact pattern on color Doppler (Fig. 10.17c). Due to anatomic and technical factors that are encountered while imaging the main renal artery (Goldberg, 1993 and Middleton, 1992) this method is very time consuming, tedious and extremely operator dependent. The wide range of peak systolic velocity values (100 to 800 cm/sec) (Avasthi et al, 1984 and rumack et al, 1991) that are used as diagnostic criteria by different authors is an indication of the unreliability of this method.

The Indirect Method

This approach is modeled after peripheral pulse changes in patients with aortic stenosis. It is technically easier to perform than the direct method. Arteries distal to the stenosis (usually the interlobar branches) are identified using color Doppler. From the spectral waveform, the acceleration time and acceleration index are electronically calculated (normal acceleration time = 0.07 seconds and acceleration index = 300 cm/sec^2). A prolonged acceleration time indicates *pulsus*

10

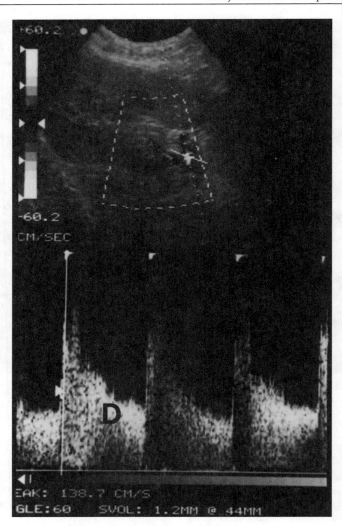

Fig. 10.17b. Spectral Doppler confirms and quantifies the high peak systolic velocity (138 cm/sec), high diastolic flow (D) and flow turbulence in both phases.

Fig. 10.17c and d. (opposite page) Perivascular tissue vibration (curved arrow) artefact on color Doppler evaluation of the main renal artery in another patient with a 60% stenosis (straight arrow). (Reprinted with permission from Nicolas Grenier et al. Detection of Vascular Complications in Renal Allografts with color Doppler Flow Imaging. Radiology 1991; 178:217-223.)

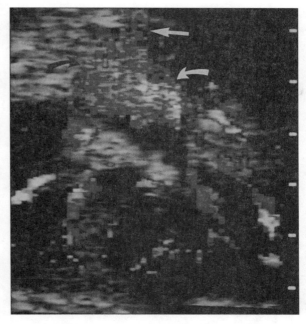

Fig. 10.17c.

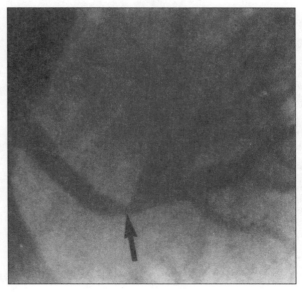

Fig. 10.17d.

tardus and a decreased acceleration index is synonymous with *pulsus parvus*. Abnormal values strongly suggest significant (> 60%) proximal stenosis (Stavros et al, 1992). Despite its seemingly positive attribution, there continues to be some debate on the value of this method (Bude et al, 1994 and Kliewer et al, 1994). Most experience with this method has been attained by studying native kidneys. The value in renal transplant artery stenosis has not yet been established, but should be equally applicable. An abnormal study, however, does not indicate the point of stenosis. For example, a transplant patient with persistent renovascular hypertension with abnormal acceleration time and acceleration index (Fig. 10.18) had a severe proximal iliac artery stenosis.

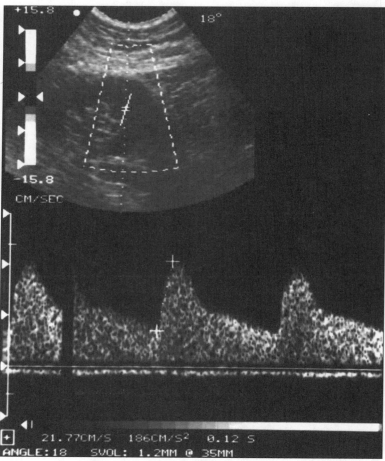

Fig. 10.18. Renal artery stenosis diagnosed using the indirect method. Spectral Doppler analysis of the interlobar arteries demonstrates absence of the early systolic peak, *pulsus tardus* and *parvus*.

Arterio-venous Fistula

Arteriovenous fistulas occasionally are seen as a complication of renal biopsy. The fistula can be detected using color Doppler and the surrounding tissue vibration can be seen (Fig. 10.19). On spectral wave analysis, a low resistant (high diastolic) flow is seen in the artery; while on the venous side, an arterial wave pattern with marked increased peak systole and turbulent diastolic flow is seen.

Pseudoaneurysm

Another rare complication of renal biopsy, the pseudoaneurysm, can be recognized as an anechoic cystic structure that demonstrates a characteristic "yin and yang" flow pattern on color Doppler (Fig. 10.20). Spectral wave analysis in the

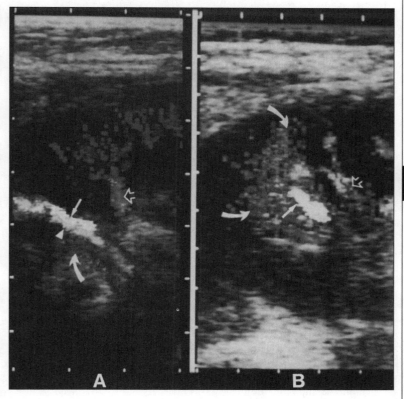

Fig. 10.19. AV fistula. A. Increased color saturation towards white (indicating high flow) is present in the artery supplying (straight arrow) and vein draining (arrow head) the fistula. The open arrow indicates a normal interlobar artery. Tissue vibration (curved arrow) is also seen. **B.** Another patient with an increased flow in the interlobar artery (straight arrow) with a normal artery for comparison (open arrow). Prominent tissue vibration (curved arrow) is present. (both from WD Middleton, GM Kellman, GL Melson, BL Medraza. Postbiopsy Renal Transplant Arteriovenous Fistulas. Color Doppler US Characteristics. Radiology 1989. 171:253-257; with permission).

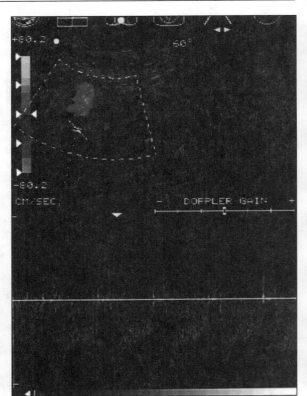

Fig. 10.20. Pseudoaneurysm. The characteristic bidirectional "yin and yang" color flow pattern in the lumen and to and fro on the spectral waveform at the neck of the aneurysm are seen.

10

neck of the aneurysm demonstrates the characteristic "to and fro" flow with reversal during diastole. Tissue vibration artefacts, such as seen in AV fistula and arterial stenosis, can also be present (Dodd et al, 1991).

Neoplasm

Any kind of renal neoplasm (e.g., renal cell carcinoma, transitional cell carcinoma, etc.) can also occur in the kidney transplant. Until proven otherwise, this diagnosis should be considered if a solid renal mass is found.

Lymphoma and lyphoproliferative disorders related to prolonged or heavy immunosuppressive therapy have the appearance of a hypoechoic mass (or masses) (Fig. 10.21). The great majority (94%) of the posttransplant lymphomas are non-Hodgkin's lymphomas (Penn, 1992); 13% of patients with this complication have either macroscopic or microscopic involvement of the allograft (Penn, 1992).

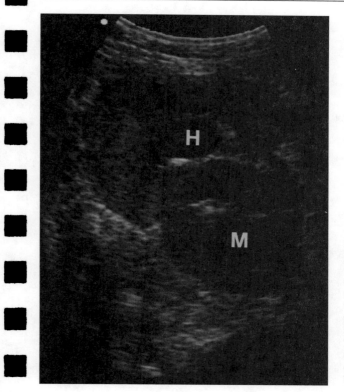

Fig. 10.21. Lympho-proliferative disorder (LPD). A hypoechoic solid mass (SM) is seen in the renal transplant hilum with secondary moderate hydronephrosis (H). The diagnosis was confirmed with a per-cutaneous CT-guided biopsy.

10

PANCREAS TRANSPLANT

The pancreas (Fig. 10.22a) appears as an elongated, echogenic and homog-enous structure cephalad and to the right of the urinary bladder. With a full blad-der, the fluid-filled duodenal segment can be seen between the head of the pan-creas and the urinary bladder (Fig. 10.22b). The main pancreatic vessels can be easily detected using color Doppler. Parenchymal perfusion can be evaluated us-ing more sensitive power color Doppler (Fig. 10.22c). The use of ultrasound in evaluation of patients with pancreas transplant is limited to detection of vascular complications (Snider et al, 1991) and fluid collections (Yuh et al, 1988) including pseudocyst. The role of resistive index and color Doppler in the diagnosis of pan-creas transplant rejection has not been established at this time.

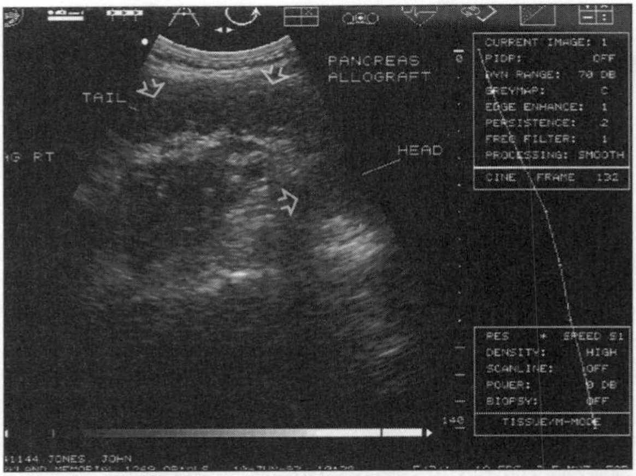

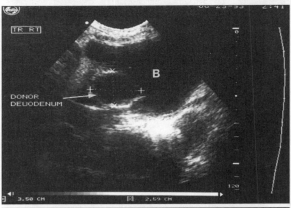

Fig. 10.22. Normal pancreas allograft. a) Normal pancreas (long axis) has a homogenous, "coarse" echogenic texture (open arrows).

Fig. 10.22b. The transplanted duodenal segment (D) can be seen especially with a full bladder (B).

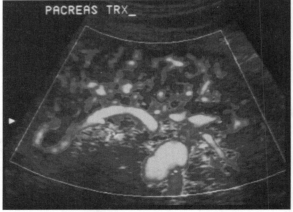

Fig. 10.22c. The pancreatic vessels are easily seen using color Doppler. This is useful in locating the pancreas since no other anatomic landmarks are present.

10

ULTRASOUND GUIDED ASPIRATION AND DRAINAGE

Ultrasound is suitable for guidance of aspiration of most fluid collections and percutaneous drainage in selected cases. After a target area has been selected, the depth, point of skin puncture and the angle of the needle tract are determined. This "blind" method is usually adequate to aspirate or drain large collections. Fluid collections that are small or close to vital structures may require continuous real time monitoring during the procedure.

A lymphocele is probably the most common fluid collection requiring aspiration or drainage. Indications for lymphocele aspiration include evaluation of a possible source of fever (infected lymphocele), to relieve renal obstruction and to evaluate the effect on renal function. If, following decompression of the lymphocele, the hydronephrosis resolves, or renal function improves and the fluid collection reoccurs, catheter drainage should be considered. This can be done using either ultrasound or CT as guidance. In our institution, CT guidance is preferred since a collapsed bowel loop in the path of the procedure can be missed on ultrasound.

SELECTED READING

1   Ellenbogen PH, Scheible W, Talner LB. Sensitivity of Gray Scale Ultrasound in Detecting Urinary Tract Obstruction. AJR 1978; 130:731-733.

2   Platt JF, Rubin JM, Ellis JH et al. Duplex Doppler Ultrasound of the Kidney: Differentiation of Obstructive from Nonobstructive Dilatation. Radiology 1989; 171:515-517.

3   Chrysi Kopolous H, Talner LB. Doppler Evaluation of Pyelocaliectasis. AJR 1990; 155:196.

4   Rigsby CM, Taylor KJW et al. Renal Allograft in Acute Rejection: Evaluation Using Duplex sonography. Radiology 1986; 158:375-378.

5   Rifkin MD, Needleman L, Pasto ME et al. Evaluation of Renal Transplant Rejection by Duplex Doppler Examination; Value of Resistive Index. AJR 1987; 148759-762.

6   Grant EG, Kaveggia LP, Perrella ER et al. Duplex Doppler Sonography in Renal Allograft: The Significance of Reversed Flow in Diastole. AJR 1990; 155:295-298.

7   Genkins SM, Sanfilippo FP, Carroll BA. Duplex Doppler Sonography in Renal Transplant: Lack of Sensitivity and Specificity in Establishing Pathologic Diagnosis. AJR 1989; 152:535-539.

8   Kelcz F, Pozniak MA, Pirsch JD, Oberly TD. Pyramidal Appearance and Resistive Index: Insensitive and nonspecific Sonographic Indicator of Renal Transplant Rejection. AJR 155:531-535.

9   Middleton WD, Erickson S, Melson GL. Perivascular color artifact: pathologic significance and appearance on color Doppler US images Radiology 1989; 171:647-652.

10   Goldberg BB. Textbook of Abdominal Ultrasound. William and Wilkins, 1993 367-380.

11   Middleton WD. Doppler US Evaluation of Renal Artery Stenosis: Past, Present and Future. Radiology 1992; 1884:307-308.

12   Avasthi PS, Voyles WF Greene ER. Noninvasive diagnosis of Renal Artery Stenosis by Echo Doppler Velocimetry. Kidney Int 1984: 25:824-829.

10

13   Rumack CM, Wilson SR, Charboneau JW. Diagnostic Ultrasound, Mosby-Year Book Inc. 1991, 254.

14   Stavros AT et al. Segmental Stenosis of the Renal Artery: Pattern Recognition of Tardus and Parvus Abnormalities with Duples Sonography. Radiology 1992; 184:487-492.

15   Bude RO, Rubin JM, Platt JF, Fechner KP, Adler RS. Pulsus Tardus: Its Cause and Potential Limitations in Detection of Arterial Stenosis: Radiology 1994; 1900:779-784.

16   Kliewer MA, Tupler RH, Hertzbert BS et al. Doppler Evaluation of Renal Artery Stenosis: Interobserver Agreement in the Interpretation of Wavefor Morphology. AJR 1994; 162:1371-1376.

17   Middleton WD, Kellman GM, Melson GL, Madrazo BL. Postbiopsy Renal Transplant Arteriovenous Fistulas: Color Doppler US Characteristics. Radiology 1989; 171:253-257.

18   Dodd III GD, Tublin ME, Shah A, Zajko AB. Imaging of Vascular Complications Associated with Renal Transplants. AJR September 1991; 157:449-459.

19   Penn I. Immunosuppression–A Contributory Factor in Lymphoma Formation. Clin Transplant 1992; 6:214-219 (Special Issue).

20   Snider JF, Hunter DW, Kuni CC, Casztaneda-Zuniga WR, Lefourneau JG. Pancreatic Transplantation: Radiologic Evaluation of Vascular Complications. Radiology 1991; 178:749-753.

21   Yuh WTC, Wiese JS, Abu-Yousef MM et al. Pancreatic Transplant Imaging. Radiology 1988; 167:679-683.

10

Urologic Complications
in Renal Transplantation

Arthur I. Sagalowsky

Surgical complications following renal transplantation may be minimized by careful attention to technique during organ procurement and subsequent engraftment. Postoperative complications may have a profound negative effect on allograft survival and on patient survival. Obviously, prevention of these complications is the best remedy. A clear understanding of the causes of complications and prompt diagnosis and treatment are the cornerstones of a successful outcome. The entire renal transplant procedure may be viewed in terms of the following component parts: the incision, the renovascular anastomosis to the iliac vessels, the ureteral reimplant and the handling of lymphatics in the renal hilum and around the iliac vessels. This chapter addresses specifically urologic complications in renal transplantation, excluding vascular complications except as they apply to the ureter and to erectile function in the male.

ANATOMIC CONSIDERATIONS

A successful transplant begins with the proper organ procurement, whether from a cadaveric or a live donor. Preservation of ureteral blood supply is critical (Fig. 11.1). The renal pelvis and proximal ureter receive arterial supply from renal hilar branches of the renal artery and directly from the aorta. Although lower pole renal arteries may appear small in their contribution to the renal parenchyma, they may be vital to the renal pelvis and proximal ureter. Thus, lower pole arteries should be preserved if possible. The pelvic portion of the ureter receives vasculature from the iliac arteries and from the side wall of the bladder. The midureter

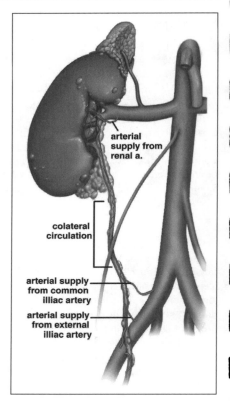

arterial
supply from
renal a.

colateral
circulation

arterial supply
from common
illiac artery

arterial supply
from external
illiac artery

Fig. 11.1. The renal pelvis and proximal ureter receive blood supply from arterial branches of the aorta and renal artery. The pelvic ureter receives blood supply from the iliac and perivesical vessels. The mid-ureter receives collateral blood supply from these proximal and distal sources.

11

does not have direct arterial supply, it receives blood from a network of vessels connecting to and from the proximal and distal sources described above. The periureteral vascular supply must be included with the ureter during dissection to protect the vulnerable midureter. Following organ procurement, the entire allograft ureter receives all of its arterial supply from the proximal renal hilar sources. Therefore, excess ureteral length should be discarded at the time of reimplantation. Use of electrocautery is avoided on periureteral vessels. Dissection within the renal hilum is minimized to avoid injury to vessels of the renal pelvis and proximal ureter.

En bloc excision is the preferred method during cadaver organ procurement. During live donor nephrectomy the proximal renal artery and the distal renal vein are isolated without dissection toward the renal hilum. The triangular area of fat between the lower pole of the kidney, the ureter and the renal hilum is undisturbed. The surgeon may mistakenly provide too short a ureter when dissection is difficult, as may occur with a large or obese donor. This problem is avoided by routinely dividing the ureter at or below where it crosses the iliac artery as a landmark.

URINE LEAK

Urine leak is a true emergency in the immunosuppressed transplant recipient. Urine in the wound is a strong irritant and predisposes to infection and disruption of the vascular anastomoses. Prompt diagnosis and cessation of the leak are essential. Urine leak is suspected clinically when the following are present: decreased urine output, increased serum creatinine, possibly increased serum chloride, increased weight, pain and fullness over the incision and fluid drainage from the wound. Ultrasonography is the usual test for identifying the fluid collection in the pelvis. Percutaneous fluid aspiration and demonstration of a creatinine concentration higher than serum creatinine confirms a urine leak. One cannot distinguish between urine leak and a lymphocele by visual inspection of the fluid.

BLADDER LEAK

The anterior cystostomy performed with intravesical ureteral reimplantation is a potential source of urine leak. The bladder is carefully closed with absorbable sutures in three layers: mucosa, muscularis and seromuscularis. In our experience at the University of Texas Southwestern Medical Center and Parkland Memorial Hospital transplant program, the cystostomy closure leak rate is only 0.6% even with early removal of the bladder catheter 24 to 48 h postoperatively. Rarely, both cystostomies leak in the paired recipients of kidneys from the same donor due to the transfer of candidal urinary infection with the allograft.

Leakage at the bladder-ureteral anastomosis with extravesical ureteral reimplantation is discussed below as a ureteral complication, rather than as a bladder problem per se.

Bladder leak may be confirmed as the specific source of a urine leak by demonstration of extravasation on cystography. The decision for management by prolonged catheter drainage vs. re-exploration and repair is based on the degree of leak and the presence or absence of infection. However, the overriding principle is that the treatment must result in prompt cessation of the leak.

URETERAL LEAK

Ureteral leak may occur from a direct injury during organ procurement, ischemic necrosis due to loss of critical blood supply during procurement, or, from mechanical or technical errors with the reimplant technique. Examples of each cause of ureteral leak will be described.

Immediate postoperative ureteral leak occurred due to a missed procurement injury in 1 of 1516 renal transplants performed at our campus over a 33-year period. At the time of transplantation, when the kidney was removed from the sterile container, the midureter was noted to be tied into the drawstring of the bag. The ureter was carefully inspected and felt to be intact. Discarding the involved segment would have left a fairly short ureter; therefore, the reimplant was performed distal to this segment. In the initial 6 h postoperatively, the patient was anuric. Physical exam revealed wound fullness and fluid drainage. The patient was returned to surgery and a pinpoint leak spurting urine from the involved midureter was noted. All the ureter distal to this point was discarded and a repeat extravesical neocystostomy was performed over a double 'J' internal stent. The reimplant healed normally and excellent graft function was obtained. In retrospect, the decision to retain the suspect ureteral segment was an error in judgment.

Ureteral leak usually is diagnosed within the first week postoperatively when the cause is a technical problem with the anastomosis. The signs and symptoms of ureteral leak are the same as described above for bladder leak. Proof of a urine leak and absence of extravasation on cystography focuses attention on the transplant ureter as the source. Ureteral leak at the anastomosis or along the tunnel did not occur in our experience with 505 consecutive intravesical ureteral reimplants. Extravesical ureteral implants are more vulnerable to leak in the perioperative period because they are not anchored to the trigone. A longer period of postoperative catheter drainage is recommended for extravesical vs. intravesical reimplants (5-7 days vs. 2-3 days). During the past 10 years at our center, two ureteral leaks occurred along the tunnel in 266 consecutive extravesical reimplants (0.8%). One leak developed on postoperative day 7 when the catheter was removed. At exploration, the ureter was still attached circumferentially to the bladder mucosa, but the ureter and anastomosis had retracted completely out of the tunnel (Fig. 11.2). This problem has been eliminated in all subsequent cases by securing the ureter at both ends of the tunnel.

Retrograde pyelography to confirm ureteral leak or to attempt stent placement should be discouraged in the postoperative period. Regional or general anesthesia is required in most patients in this setting and the procedure has a low success rate with a fresh ureteral reimplant. The attempt at retrograde stent placement may further damage the anastomosis. Percutaneous placement of a nephrostomy is the preferred procedure. Successful access to the collecting system almost always is possible by the skilled interventional radiologist. This provides proximal renal drainage and allows accurate diagnosis via an antegrade nephroureterogram. Small ureteral leaks due to technical problems at the tunnel or anastomosis may be managed definitively by placement of an antegrade stent provided that contrast passes into the bladder, demonstrating continuity of the system and providing a flexible guidewire can be passed across the anastomosis (Fig. 11.3). However, percutaneous management is not recommended if ureteral necrosis is suspected (see below).

Fig. 11.2. (opposite page) a) Ureteral leak in extravesical reimplant due to retraction of the intact anastomosis through the ureteral tunnel. b) The ureteral reimplant is reestablished. The ureter is secured at the proximal end of the tunnel and a double J stent is inserted.

Fig. 11.3. (opposite page) Mild ureteral leak along the tunnel site with an intact anastomosis (a) and healthy ureter may be managed simply by insertion of a double J stent (b).

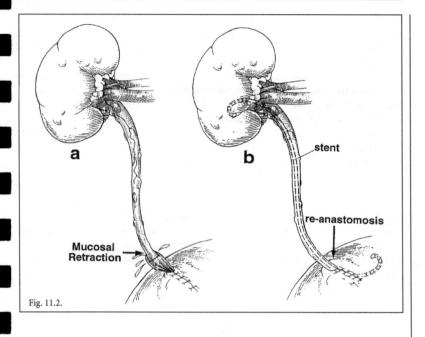

Fig. 11.2.

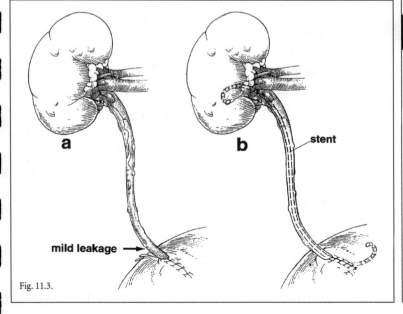

Fig. 11.3.

11

Ureteral leak due to ischemic necrosis of a segment of ureter is the most diffi-cult problem. Initially postoperatively, the ischemic ureter has poor peristalsis and allograft dysfunction due to hydronephrosis which may be demonstrated by ul-trasonography or radionuclide scan. Finally, by 7 to 21 days postoperatively, ure-teral necrosis occurs and the signs of urine leak and possible infection are seen. A percutaneous antegrade nephrostogram will reveal ureteral irregularity ("scallop-ing") and a large amount of extravasation. Ureteral continuity with contrast pas-sage into the bladder may or may not be present.

Prompt surgical exploration is strongly advocated in these cases. Debridement, excision and repeat ureteral reimplantation may be attempted if only the distal ureteral segment is affected and if there is no sign of infection (Fig 11.4). How-ever, in most instances, a considerable length of ureter is necrotic and local infec-tion is a major concern. Debridement and excision back to healthy ureter or renal pelvis is essential. If the bladder is mobile, one may consider pyelovesicostomy in the absence of infection. Nevertheless, the author believes that the safest approach is placement of a nephrostomy, careful closure of the proximal collecting system and establishment of closed suction wound drainage (Fig 11.5). This course of action provides maximal control over urinary drainage and minimizes the risk of abscess formation in this difficult setting. A delayed repair, 3 to 6 months later by ureteropyelostomy of the ipsilateral native ureter anastomosed to the transplant

11

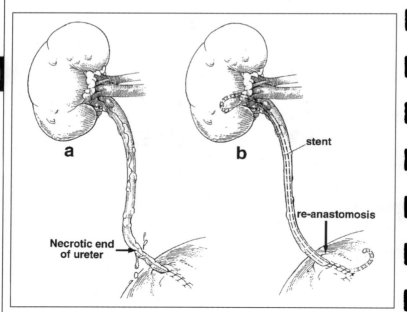

Fig. 11.4. a) Ureteral leak due to short necrotic area of distal ureter at the reimplant site. b) Repair by resection of necrotic segment, re-do ureteral anastomosis, and placement of an internal double J stent.

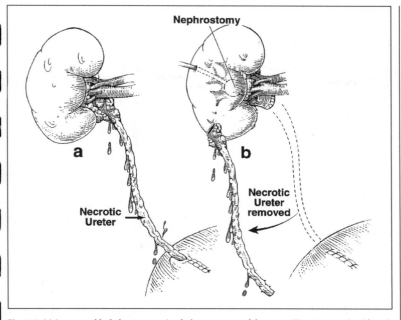

Fig. 11.5. Major ureteral leak due to necrosis of a long segment of the ureter. First stage repair with excision of the entire necrotic ureter; closure of the renal pelvis and placement of a nephrostomy tube.

11

renal pelvis has a high success rate (Figs. 11.6, 11.7). Ipsilateral native nephrectomy is performed at the same time to prevent silent obstruction and abscess formation in the native kidney.

URETERAL OBSTRUCTION

Transient acute obstruction due to edema along the ureteral tunnel always is a concern when the initial postoperative urine output is intermittent. However, the problem almost always is self-limiting if proper technique for the ureteral reimplant is followed. The ureter must be free to peristalse within the tunnel. The ureter must not be angulated at the entry site into the tunnel. These technical features are necessary for both intravesical and extravesical implants.

Chronic ureteral obstruction develops in some cases due to ischemia and fibrosis. The possible role of chronic rejection in ureteral fibrosis is speculative. Progressive rise in serum creatinine not due to acute rejection and the development of hydronephrosis suggests the diagnosis. Retrograde pyelography and stent placement continue to have a role in the setting of chronic ureteral obstruction.

Fig. 11.6. Continuity of urinary drainage to the bladder is re-established by a delayed secondary procedure consisting of ipsilateral native nephrectomy and ureteropyelostomy of the native ureter to the allograft renal pelvis.

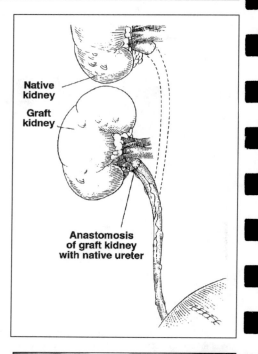

Fig. 11.7a. Nephrostogram of healed renal pelvis after complete excision of necrotic ureter as in Figure 5a.

11

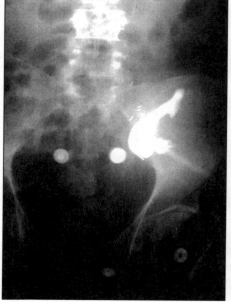

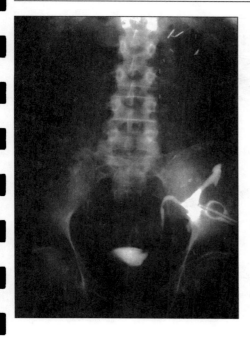

Fig. 11.7b. Nephrostogram after ureter-pyelostomy with native ureter as in Figure 6.

However, percutaneous antegrade approaches to nephrostomy and stent placement are increasingly favored. A soft indwelling stent is placed for 2 to 4 weeks in most cases. A variety of endoscopic procedures may provide definitive repair. Balloon dilation alone is the least invasive technique and succeeds in 20-50% of short ureteral strictures (Fig 11.8). Direct visual incision is possible with endoscopic scissors or with an electrosurgical wire or laser. Longer ureteral strictures may be managed with the Acucise cutting wire balloon (Fig 11.9). The elongated balloon has an electrocautery wire running lengthwise, which cuts through the stricture when activated. The balloon is placed across the stricture under fluoroscopy. Precise three-dimensional orientation is required to orient the cutting wire away from major blood vessels. Following any of the above procedures, a ureteral stent is left for 2-6 weeks.

At times it is difficult to distinguish renal allograft dysfunction due to ureteral obstruction which needs treatment from the new appearance of mild hydronephrosis which is merely a component of far advanced allograft dysfunction associated with chronic rejection. In the late stages of chronic rejection, the renal collecting system may have poor peristalsis and demonstrate mild hydronephrosis. Failure to appreciate this fact results in invasive procedures such as nephrostomy and stent placements which will not benefit the patient.

The spermatic cord may be a source of external ureteral obstruction. The ureter is placed under the spermatic cord at the time of ureteral reimplantation to

11

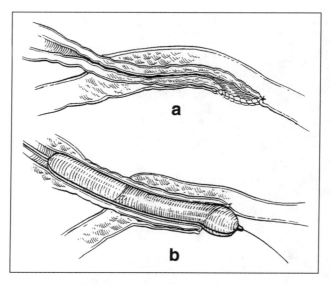

Fig. 11.8. Balloon dilation of the ureterovesical stricture.

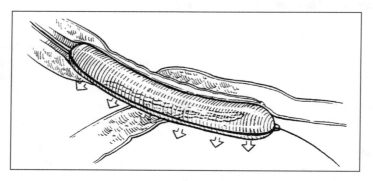

Fig. 11.9. Definitive incision of ureteral stricture with the Acucise cutting balloon used under fluoroscopic control.

prevent this problem (Fig 11.10). A small kidney may rotate in the wound and produce either renovascular thrombosis or torsion of the ureter around the spermatic cord (Figs. 11.11, 11.12). Nephropexy of small kidneys prevents this problem.

LYMPHOCELE

The lymphatics surrounding the iliac vessels should be carefully tied or clipped during exposure of the vessels for anastomosis of the renal artery and vein to

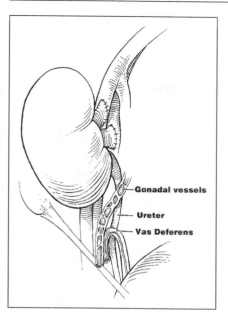

Fig. 11.10. The ureter must be placed under the spermatic cord for the ureteral reimplantation to the bladder.

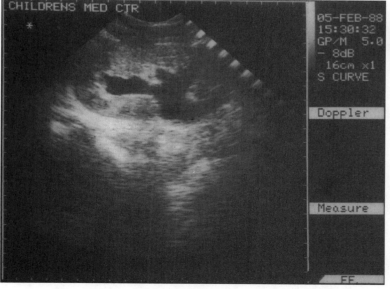

Fig. 11.11. Ultrasound reveals marked hydronephrosis in a patient with anuria following transplantation of a pediatric renal allograft.

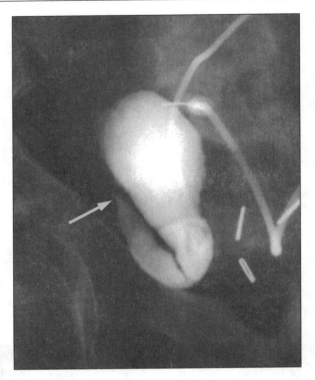

Fig. 11.12. Antegrade nephrostogram in the same patient as Figure 11 reveals a markedly dilated ureter tapering to complete obstruction at the very distal ureter. Subsequent surgical exploration revealed torsion of the allograft such that the allograft ureter was wrapped around and obstructed by the spermatic cord even though the ureter was correctly positioned beneath the spermatic cord for the ureteral reimplant.

minimize the risk of postoperative lymphocele formation. Some transplant surgeons advocate dissection and exposure of the minimum length of the iliac vessels which permits anastomosis of the renal vessels to decrease the risk of lymphocele formation. In the authors' experience, complete mobilization of the external iliac artery and vein greatly facilitates performance of the renovascular anastomoses and is not associated with an increased risk of lymphocele provided the lymphatics are meticulously secured. A degree of lymph leak from divided renal hilar lymphatics also may contribute. However, studies with radio-labeled tracers demonstrate that the major source of the lymph in lymphoceles is from the iliac lymphatics. Most lymphoceles appear in the first several months postoperatively. This fits with the proposed etiology. The cause of occasional late appearance of lymphoceles is unclear.

Small asymptomatic lymphoceles are common and are of no consequence. The signs and symptoms of a lymphocele may include increased serum creatinine due to extrinsic compression of the ureter, edema of the ipsilateral lower extremity or pain and fullness over the wound and lower abdomen. Diagnosis is established by the finding of a fluid collection on ultrasonography and percutaneous aspiration demonstrating that the fluid is lymph and not urine (i.e., fluid creatinine and potassium concentrations similar to serum). Fever and turbid fluid suggest infection of a lymphocele.

The size of a lymphocele may bear no relationship to the symptoms produced. The author has seen a patient present with a large lower abdominal mass appearing like a pregnancy at term due to a lymphocele that contained three liters of fluid without any hydronephrosis or leg swelling present (Fig. 11.13). At the other extreme, a 2-3 cm lymphocele with a thick fibrous rim may produce severe extrinsic compression of the external iliac vein or ureter and produce marked swelling of the lower extremity or hydronephrosis of the allograft.

The initial diagnostic aspiration and drainage may provide effective treatment in a minority of cases. However, most lymphoceles recur and require definitive treatment. Laparoscopic intraperitonealization of the lymphocele has become the preferred treatment (Fig.11.14). A large window is created between the medial wall of the lymphocele and the peritoneal cavity so that further lymphatic leak is reabsorbed. Ultimately, the lymphocele cavity collapses and the lymphatic leak

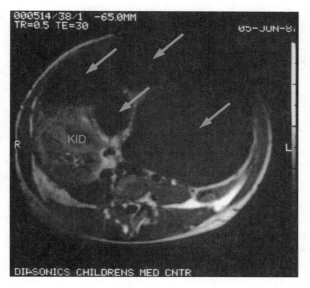

11

Fig. 11.13. MRI demonstrating 3 liter multiloculated pelvic lymphocele in renal transplant patient without ureteral obstruction or lower extremity edema.

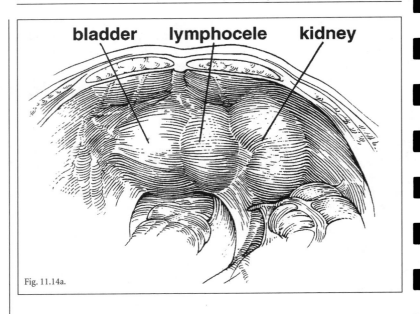

bladder   lymphocele   kidney

Fig. 11.14a.

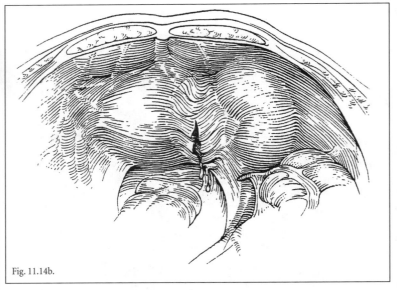

Fig. 11.14b.

Fig. 11.14. A laparoscopic view into the pelvis revealing lymphocele between the bladder and renal allograft. b) A large opening is created on the lymphocele to allow intraperitoneal drainage of the lymphatic fluid.

points sclerose. Open surgical intraperitoneal drainage and prolonged percutaneous closed tube drainage also are effective alternative treatments. The latter procedure usually requires 3-6 weeks of tube drainage and runs the risk of introducing infection. Infected lymphoceles must be treated by external drainage (open or via closed tubes) and antibiotics. Repeated instillation of sclerosing agents, i.e., povidone, has been reported as being highly effective.

HYDROCELE

The spermatic cord is mobilized and preserved in male recipients whenever possible. In a few patients, exposure to the iliac vessels is compromised by a tightly bound spermatic cord due to prior pelvic surgery and scarring or due to extreme obesity. In these cases, the spermatic cord is ligated and divided to increase the safety of the transplant procedure. Acute division of the spermatic cord will result in hydrocele formation in approximately 50% of cases and testicular atrophy in a variable number. Persistent symptomatic hydroceles are treated by standard surgical hydrocelectomy.

ERECTILE DYSFUNCTION

Satisfactory erectile function in males requires both systemic hormonal and psychologic contributions and local neurovascular input. Many patients with end stage renal disease have organic impotence due to vasculogenic and/or neurogenic causes due to their primary disease process. The arterial supply to the penis comes from the internal pudendal arteries which are branches of the internal iliacs. Male recipients are asked about the status of erectile function preoperatively. If erectile function is present but diminished, anastomosis of the renal artery to the external iliac rather than to the internal iliac is preferred to preserve penile arterial supply. In repeat male transplant recipients, it is important to know if the contralateral internal iliac artery was utilized with the prior transplant.

11

SUMMARY

Surgical complications overall and urologic complications in particular, occur in only a small percentage of renal transplants. Prompt diagnosis and treatment of these complications allows for minimizing the impact on graft and patient survival.

Immunosuppressive Drugs in Kidney and Kidney/Pancreas Transplantation and Related Complications

Lucille A. LoTempio, Carolyn E. Munschauer

INTRODUCTION

The first successful renal transplant was performed in 1954, but it wasn't until the 1960s that azathioprine and polyclonal antilymphocyte preparations were employed. The introduction of cyclosporine (CsA) in 1984 was a major turning point in the pharmacotherapy of transplantation which markedly improved patient and graft survival. Pharmacologic factors contributing to the success of transplantation in the 1980s included management of cytomegalovirus (CMV) infection with ganciclovir and the approval of Muromonab-CD3 (OKT3®, Ortho-Biotech). With the recent incorporation of tacrolimus, mycophenolate mofetil and cyclosporine microemulsion, Neoral® (Novartis Pharmaceuticals), and SangCya™ (SangStat) into the armamentarium of immunosuppressive agents, one-year graft survival is 87% and 93% for recipients of cadaveric donor renal transplants and living donor renal transplants, respectively. Long term graft survival (greater than 1 year) has also improved since the introduction of CsA. It is anticipated that the use of CsA preparations with more consistent drug exposure and less interpatient variability, will positively influence the observed half-life of the organ. Also in the last year (1998) several new immunosuppressive agents have gained FDA approval or entered clinical trial (Table 12.1). These include drugs effective against T and B cells, at several stages of posttransplant immune activation: monoclonal and polyclonal antibody formulations; antimetabolites (both pyrimidine and purine biosynthesis inhibitors); inhibitors of lymphocyte maturation, and a new cyclosporine preparation (Table 12.1). With increased understanding of the immune system, including tolerance induction, and the development of new immunosuppressive agents, the future offers a very promising outlook for transplant recipients.

12

Kidney and Pancreas Transplantation, 2nd Edition, edited by Ingemar J.A. Davidson © 1999 Landes Bioscience

Table 12.1. Immunosuppression drug time line

Year	Drug introduced/Event
1954	First kidney transplant
1961	Azathioprine
1980	ALG
1981	Atgam
1983	Sandimmune
1987	OKT3
1994	Tacrolimus
1995	Neoral
1995	CellCept
1997	Thymoglobulin
1998	Sang CyA
1998	Zenapax
1998	Rapamune
1999	Leflunomide, and others in trials

IMMUNOSUPPRESSIVE AGENTS

NUCLEOTIDE SYNTHESIS INHIBITORS

Early immunosuppression consisted of cytotoxic agents such as azathioprine (AZA) or cyclophosphamide, which target proliferating myelocytes after antigenic stimulation through interference with nucleic acid metabolism. AZA is still used for the prevention of renal allograft rejection in combination with CsA and corticosteroids. Mycophenolate mofetil (MMF, CellCept,® Roche) blocks the de novo pathway of purine biosynthesis and may specifically interfere with the recruitment of lymphocytes and monocytes to sites of inflammation. It has replaced azathioprine as part of triple drug therapy in many centers, largely due to its selective effect on lymphocytes. Multicenter trials including over 1400 renal transplant patients receiving MMF vs AZA or placebo in conjunction with CsA and steroids revealed a statistically significant reduction in the number of biopsy proven rejections at one year in the MMF treated patients. Mizorbine (Bredinin©, Asahi Chemical Industry Co. of Japan), a purine biosynthesis inhibitor currently in use in Japan, has similar efficacy to azathioprine, but has demonstrated less myelo- and hepatotoxicity. Brequinar Sodium (DuPont Merck Pharmaceutical Co., Wilmington, DE) and Leflunomide (Arava©, Hoechst Marion Roussel), both pyrimidine biosythesis inhibitors, deplete circulating bases, diminishing RNA and DNA synthesis, and therefore, cellular proliferation. In early studies Brequinar Sodium was associated with thrombocytopenia. Leflunomide may also inhibit tyrosine kinase activity and T cell cytokine response in the IL-2 signal transduction pathway. Leflunomide has exhibited remarkably few side effects in clinical trials and in treatment of rheumatoid arthritis (Table 12.2).

CYTOKINE TRANSCRIPTION INHIBITORS

Unlike the cytotoxic drugs, cyclosporine (CsA) and tacrolimus (TL) exert their effects without killing effector cells, and are not effective once antigenic stimulation has occurred. CsA and TL inhibit IL-2 elaboration by stimulated helper T cells and prevent differentiation into cytotoxic T cells. The specificity of these agents has revolutionized immunosuppressive therapy, in that they do not alter preformed immunity and the humoral immune system, minimizing infectious complications. Neoral®, introduced in 1995, has greater bioavailability over the dosing interval than cyclosporine A. SangCya™ (SangStat), available in 1998, demonstrated bioequivalence to Neoral, but in a liquid form. Tacrolimus (FK506, Prograf®, Fujisawa USA) is more potent than CsA on a weight basis, but is equivalent to CsA in action. It has also been effective as rescue therapy in kidney transplant patients with refractory rejection (Table 12.3).

CORTICOSTEROIDS
    Corticosteroids block primary immune response through inhibition of cytokine expression, including IL-1, -2, -3,-6; tumor necrosis factor (TNF) and gamma interferon. Anti-inflammatory effects include inhibition of monocytes to areas of inflammation and blockage of chemoattractant agents, permeability-increasing agents and vasodilators. Corticosteroids have been the mainstay of immunosuppression for many years, and continue to be first line treatment of rejection and primary prophylaxis in combination with CsA, TL, AZA or MMF (Table 12.4).

ANTI-T CELL ANTIBODIES AND OTHER INDUCTION THERAPIES: POLYCLONAL & MONOCLONAL ANTIBODIES, CELLULAR ANTAGONISTS
    Polyclonal antilymphocyte antibodies or monoclonal antibodies are used as immunosuppressive induction therapy in the immediate posttransplant period. Their mechanisms of action involve a reduction in the number of circulating T lymphocytes. Polyclonal antobodies, previously comprised of Minnesota ALG and Atgam, now include Thymoglobulin® (SangStat, Menlo Park, CA), a rabbit anti-thymocyte globulin (Table 12.1). In studies comparing Thymoglobulin to Atgam in episodes of rejection, Thymoglobulin is reported to be at least as effective as Atgam in reversing acute rejection episodes. This finding may provide an alternative polyclonal antibody formulation to patients sensitive to Atgam or other equine exposures (Table 12.5). OKT3 is a murine monoclonal antibody that has been available for several years. Recently, several humanized monoclonal antibodies have been introduced. The inherent advantage of a humanized antibody is loss of human anti-mouse antibody reaction. Basiliximab (Simulect®, Novartis), a chimeric anti-IL2 receptor monclonal antibody, is uniquely cell-specific, currently administered in two 20 mg doses, one before transplant and a second 3-4 days posttransplant. Daclizumab (Zenapax®, Roche Pharmaceuticals), is also an anti-IL2 receptor monoclonal, and is anti-Tac (part of the IL-2 receptor complex expressed only on activated lymphocytes). Zenapax given (1mg/kg) in five doses over an eight week period. The first dose is given pretransplant, with each subsequent dose given at a 14 day interval. Studies have shown that with this dosing regimen, Tac expression was blocked for 120 days (Table 12.6). Comparative studies suggest similar graft survival and incidence of rejection episodes with monoclonal and polyclonal antibodies.

12

Table 12.2. Pertinent features of nucleotide synthesis inhibitors

	Azathioprine (AZA) Imuran®	Mycophenolate Mofetil (MMF) CellCept®	Mizorbine (Bredinin®)	Brequinar Sodium	Leflunomide (Arava®)
Phase	Maintenance	Maintenance	Maintenance	Maintenance	Maintenance
How Supplied	Tablets 50 mg IV 100 mg/ml	Capsules 250 mg, 500 mg	Still in trials	Trials suspended	Still in trials
Dose	1-2 mg/kg once daily	1 -1.5 g twice daily	Still in trials	Trials suspended	Still in trials
Duration	Continuous	Continuous	Continuous	Continuous	Continuous
Monitoring	WBC Count	WBC Count	WBC Count	WBC Count	WBC Count
Mechanism of Action	Interfere w/nucleic acid metabolism	de novo purine biosynthesis	purine biosynthesis	pyrimidine biosynthesis	pyrimidine biosynthesis
Site of Action	Myelocytes	Lymphocytes	Myelocytes	Lymphocytes	IL-2 pathway
Adverse Effects	Dose-related bone marrow suppression. A reduction in dose is suggested for patients with WBC < 4000/mm3 and discontinued for WBC < 3000/mm3	Diarrhea, dyspepsia and vomiting	Minimal in trials	Significant thrombocytopenia in trials	Minimal in trials
Drug Interactions	Allopurinol	Antacids and cholestyramine	Not yet identified	Trials suspended	Not yet identified

12

Table 12.3. Pertinent features of cytokine transcription inhibitors

	Cyclosporine (CsA) Sandimmune®	CsA Microemulsion Neoral®	CsA Oral Solution SangCya®	Tacrolimus (FK506) Prograf®
Phase	Induction/maintenance/rejection	Same	Same	Same
How Supplied	25, 50, 100 mg capsules 100 mg/ml solution	25 & 100 mg capsules 100 mg/ml solution IV- 50 mg/ml	100 mg/ml solution	1 and 5 mg capsules IV-5 mg/ml
Dose	Oral range 4–12 mg/kg/day in 2 divided doses IV dose 1/3 of oral dose administered over 24 h	Same	Same	Oral range 0.15–0.3 mg/kg/day in 2 divided doses IV same as CsA
Duration	Continuous	Continuous	Continuous	Continuous
Monitoring	12 h trough levels. For blood levels, see chapter 6	Same	Same	Same
Assay	TDx/TDxFLx® CsA Monoclonal Whole Blood Assay, Abbott Labs			IMX® Tacrolimus Assay, Abbott Labs
Mechanism of Action	Forms complex with receptor protein, binds to calcineurin, impairs calcineurin ability to move activated T cell proteins across cell membrane.			Also complexes with specific protein competes w/calcineurin binding site.
Site of Action	Cytoplasm			Same
Adverse Effects	Nephrotoxicity, neurotoxicity (seizures) cosmetic effects: hirsutism, gingival hypertrophy			Same no cosmetic effects
Drug Interactions:	**increase blood level** ketoconazole, fluconazole, erythromycin, clarithromycin, diltiazem, verapamil, nicardipine, methylprednisolone **decrease blood level** phenobarbital, phenytoin, rifampin, nafcillin **increase toxicity** acyclovir, aminoglycosides, ganciclovir, cimetidine, NSAIDS,ACE inhibitors			Same Same Same

12

Table 12.4. Corticosteroids: Highlights of corticosteroids used in transplantation

	Prednisone Deltasone® and various others	Methylprednisolone Sodium Succinate SoluMedrol®, others
Phase	Induction/maintenance/rejection	
How Supplied	prednisone–1, 5, 10 and 20 mg tablets methylprednisolone–IV 40, 125, 500 and 1000 mg	
Dose	prednisone: chapter 6, Table 6.13. Once-daily dosing is adequate methylprednisolone IV 250 mg–1000 mg; over 3–5 days for acute rejection; See chapter 6.	
Duration	Continuous over first 24 months posttransplant, see chapter 6 for details on steroid taper and long range steroid plan.	
Monitoring	No blood or tissue levels, dose taper used to avoid marked side effects	
Assay	None	
Mechanism of Action	Inhibition of cytokine gene expression, generalized lymphopenia	
Site of Action	Cytoplasm	
Adverse Effects:	The adverse effects are well known; For the multitude of side effects, the reader is referred to any internal medicine text	
Drug Interactions	Phenytoin, barbiturates, rifampin decrease plasma levels. Ketoconazole, oral contraceptives increase plasma levels.	

12

OTHER INDUCTION THERAPIES

CELLULAR ANTAGONISTS

Another strategy employed in recent induction drugs has been inhibition of lymphocyte maturation. In studies combining 15-deoxyspergualin (DSG, commercially used in Japan) with corticosteroids, circulating CD8- cells and IL2 receptor + cells were decreased in allograft biopsy specimens, while there was an overall increase in nonspecific suppressor T cells in the same specimens. Finally, Sirolimus (Rapamune®, Wyeth-Ayerst Laboratories) is a macrolide antibiotic used in conjunction with CsA and steroids. It inhibits the binding of transcription factors induced by IL-2, and diminishes the capability of cytokines to signal T cells to enter cell division. It has synergistic activity when combined with CsA, with some patients reportedly able to discontinue steroid use (Table 12.7).

Table 12.5. Polyclonal Anti-T-cell antibodies

	Anti-Thymocyte Globulin (Equine) (ATG) Atgam®	Anti-Thymocyte Globulin (Rabbit) Thymoglobulin®
Phase	Induction/rejection	Rejection
How Supplied	IV 50 mg equine IgG per ml	IV 5 mg rabbit IgG per ml
Safety/Precaution	Preuse skin test (5 mcg)	None established
Dose	IV dose 10–15 mg/kg/day Infused over 4 h For details see chapter 6, Table 6.11	1.5 mg/kg/day infused over 4-6 h
Duration	5-14 days immediately posttransplant	7-14 days for acute rejection
Monitoring	Total lymphocyte, WBC, platelets during therapy	Same
Assay	None	None
Mechanism of Action	Widespread T cell lysis, possible surface antigen inhibition	Same
Site of Action Adverse Effects	Cell surface Hypersensitivity/anaphylaxis in < 1% of patients Serum sickness,thrombocytopenia and leukopenia	Use of rabbit IgG lessens incidence of murine reactivity Same.
Drug Interactions	Possibility for immunosuppression if maintenance protocol is continued	

12

COMPLICATIONS SECONDARY TO IMMUNOSUPPRESSION
Infection

Infections in transplant recipients can be stratified along a time continuum over the first months following the transplant procedure (Fig. 12.1). The infections prevalent during these time frames closely parallel changes in immunosuppression regimens commonly used.

In the first month following the transplant procedure, the most common infections involve the urinary tract, respiratory system, surgical wound or bacteremia from indwelling lines or catheters. The incidence of these infections, similar to general surgical patients, increases with prolonged hospitalization, intubation, repeat surgical explorations, and other complications of the operative procedure.

Table 12.6. Monoclonal Anti-T-cell antibodies

	Muromonab-CD3 (Murine) Orthoclone OKT3®	Basiliximab (Simulect®)	Daclizumab (Zenapax®)
Phase	Induction/rejection	Induction	Induction
How Supplied	5 mg/ 5ml IV	25 mg/ 5 ml	
Dose	5 mg IV once a day infused as bolus over 5 min. For details see Chapter 6.	20 mg IV infused on 2 separate days	1 mg/kg infused 5 times over 8 weeks
Duration	5-14 days immediately posttransplant for induction 5-7 days for rejection	First dose pretransplant Second dose day 4 posttransplant	First dose within 24 h of transplant. 4 additional doses each 2 weeks apart
Monitoring	Increased number of circulating CD3+ T cells indication of possible anti-OKT3 antibodies	Clinical monitoring of graft function	Clinical monitoring of graft function
Assay	Anti-OKT3 ab, or CD3+ cell count by flow cytometry	None	None
Mechanism of Action	Inactivates CD3 complex via binding to CD3 subunit (part of TCR), causes TCR inactivation on cell surface.	Anti IL-2r, prevents IL-2 upregulation	Binds to Tac subunit of IL-2r (only expressed on activated lymphocytes)
Site of Action	T cell receptor complex	IL-2r surface	IL-2r surface
Adverse Effects	Anaphylaxis (bronchospasm), worse if fluid overload is present. Cytokine release syndrome (CRS) with first dose (fever, chills, malaise, nausea and vomiting).	Humanized preparations virtually eliminate side effects associated with anti-murine abs.	Possible anaphylaxis with repeat administration. No CRS observed.
Drug Interactions	Possibility for over-immunosuppression if maintenance protocol is continued during treatment of rejection.		

12

Table 12.7. Lymphocyte maturation inhibitors, cytokine signalling antagonists

	15-Deoxyspergualin (DSG)	Sirolimus (Rapamune®)
Phase	Induction/rejection	Same
How Supplied	Still in trials	Still in trials
Dose	3-5 mg/kg/day in slow IV infusion	2-5 mg concomitant w/CsA
Duration	7-10 days for induction and acute rejection	Long term
Monitoring	Still in trials	Still in trials
Assay	None	None
Mechanism of Action	Inhibition of lymphocyte maturation, does not alter lymphocyte proliferation; may impede production of antigen-specific T cells.	Similar to CsA and Tacrolimus, but does not inhibit calcineurin. Impairs cytokine signalling of T cells.
Site of Action	Unclear presently—possibly late in T cell activation process.	Cytoplasm
Adverse Effects	Early reports of bone marrow suppression, paresthesias, hypotension during infusion, GI disturbances.	Thrombocytopenia, leukopenia, hyper-cholesterolemia
Drug Interactions	Still in trials	FK506(binding site competition)

12

Less common are active infections transferred with the graft and superinfection of patients transplanted in spite of an active infection. Rarely seen in this time period are opportunistic infections (i.e., *Listeria moncytogenes*, aspergillus). If these pathogens are present early following transplant, it may be an indication of noso-comial contamination, existence of immunosuppression before transplantation, or transfer from donor to recipient.

Immunomodulating viral infections are usually not present during this early period, despite the highest levels of immunosuppression most patients will face. The net state of immunosuppression is measured over time from pretransplant throughout each patient's posttransplant course. It is a valuable factor in evaluating the likelihood of infection and rejection and must be adjusted regularly over the time course of transplantation. This consists of the interaction between im-munosuppressive therapy (dose, duration, choice of agents), viral status and viral history of the recipient and donor (i.e., CMV status) and early complications arising from the transplant procedure.

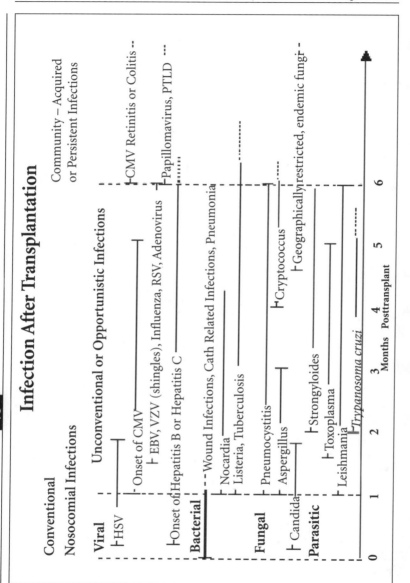

Fig. 12.1. Timeline of common infections following renal transplantation. Modified with permission from the American Journal of Medicine 1981, 70:405-411.

The next focal period for posttransplant infection is from one to six months. For most patients, the immunosuppression regimen is now chronic and sustained, resulting in maximal immunomodulation. This allows primary infection or reactivation of viral pathogens, including CMV, EBV, herpesviruses, hepatitis B or C and HIV, if present. Additionally, if a patient experienced early infectious complications, including nosocomial pathogens or surgical complications, these are likely to recur or remain chronic during this period. If the immunosuppressive state is maintained at a high rate during this period, patients may experience additional opportunistic infections, including L. monocytogenes, *Pneumocystis carinii* and aspergillus.

The pattern of infection changes after six months posttransplant. At this point, immunosuppression is adjusted depending on individual patient graft function and postoperative course, including infection profile. Patients can now be divided into three categories.

The majority (> 80%) have acceptable organ function without rejection or multiple toxicity episodes, and minimal infectious complications following an uncomplicated surgical procedure. In these patients, immunosuppression can be systematically reduced to achieve a balance between optimal organ function and immune system competence to invading pathogens. The long term infection profile in this group parallels the general population, with most instances of infection respiratory in nature. The prevalence of opportunistic infections in these patients is unusual, but can occur in the presence of an unusual exposure.

Second, approximately 5-10% of transplant recipients have repeat, early rejection episodes or develop chronic rejection. These result in higher levels of immunosuppressive medications, and often the use of several classes of immunosuppressives in a short time period. These patients not only suffer from the metabolic consequences of suboptimal organ function, but are also predisposed to infectious complications, most notably from opportunistic agents and viral pathogens.

Finally, up to 10% of organ recipients will have progressive viral infection, most often CMV, EBV or hepatitis B or C. In these patients, the chronic viral status may cause damage to the allograft or other organs necessitating reduction or cessation of immunosuppression. Some viral processes may lead to malignancy, vascular or neurologic pathologies, end organ disease or death.

With the identification of an infectious profile in the transplant population, antimicrobial therapy has moved away from therapeutic treatment of established infection to prophylaxis and preemptive therapy. The hallmark viral and opportunistic pathogens exposed to this theory have been *P. carinii* pneumonia (PCP), CMV, candidiasis and aspergillus.

Cytomegalovirus (CMV) is of special importance in transplantation because it is transferable from the donor allograft, can be reactivated in the recipient or occur as a superinfection, and is released in response to common transplant sequelae, including antilymphocyte antibodies, cytotoxicity, systemic infection, and inflammatory processes. Once reactivated, the virus replicates in a cell specific manner. The usual host defense against CMV are MHC-associated, viral specific

12

T lymphocytes, which are generally suppressed in the transplant recipient by antilymphocyte antibody preparations. In addition, cytokine transcription inhibitors such as cyclosporine and tacrolimus function in suppression of the T cell response, thus disabling the host antiviral defense. Clinical infection appears as mild flu-like symptoms, prolonged fevers, malaise, leukopenia, pneumonia, hepatitis, gastrointestinal ulcers and retinitis. Laboratory signs include leukopenia, thrombocytopenia and lymphocytosis. CMV viral replication occurs in numerous cell types, and upregulates production of cytokines, adhesion molecules, oncogenes and and other products of DNA and RNA synthesis. The end result of this cellular cascade is end organ damage, tissue invasion and MHC class II antigen presentation in the allograft. Populations at risk for CMV superinfection or reactivation are the seronegative recipients of seropositive donors or seropositive recipients receiving higher than usual immunosuppression. Prevention of CMV infection is accomplished by careful pre-emptive screening and prophylaxis with immunoglobulin preparations including hyperimmune globulin (Cytogam®), carefully tailored to the recipient and donor viral status and recipient rejection history. The ganciclovir dose is titrated against the level of renal function (Table 12.8).

It has been the author's experience to utilize oral acyclovir 400 mg three times daily for 12 weeks as prophylaxis against numerous viral pathogens. Low dose acyclovir has been shown to be effective in preventing reactivation of herpes simplex stomatitis. The dose of acyclovir is reduced to 200 mg twice daily in patients with delayed graft function. High serologic risk patients receive IV ganciclovir prophylaxis posttransplant for approximately 10 days, followed by oral acyclovir as stated above.

An effective prophylaxis against PCP is sulfamethoxazole/trimethoprim (Bactrim) administered as a single strength tablet daily. Bactrim also decreases the incidence of many other posttransplant bacterial infections, including upper respiratory and urinary tract infections. Toxoplasmosis as a complication of Bactrim administration is uncommon in visceral transplant recipients. Prophylaxis against superficial fungal infections such as oropharyngeal candidiasis is achieved with early administration of nystatin oral suspension 500,000 units swish and swallow three times daily. Deep fungal infections are uncommon in renal transplant recipients with an uncomplicated surgical course.

CENTRAL NERVOUS SYSTEM INFECTION

The most common manifestations of CNS infection are primarily related to opportunistic and viral infections. Acute meningitis is usualy secondary to Listeria infection, while chronic meningitis may be Cryptococcal in origin or related to systemic infection with Listeria or other pathogens. It can also be secondary to posttransplant lymphoproliferative disease (PTLD) associated with EBV. Brain infection presenting as seizures and focal deficits are usually associated with aspergillus, but can also occur as the result of EBV induced PTLD. Dementia of a progressive nature may be viral (CMV, EBV, herpes simplex) or possibly related to toxic effects of cytokine transcription inhibitors, which include demyelination.

Table 12.8 Ganciclovir doses adjusted for renal impairment

Creatinine clearance (milliliters/minute)	Ganciclovir dose (milligrams/kilogram)	Dosing interval (hours)
> 70	5	12
50–69	5	24
25–49	2.5	24
10–24	1.25	24
< 10	1.25	3 times per week following each hemodialysis session

POSTTRANSPLANT LYMPHOPROLIFERATIVE DISEASE (PTLD)

Epstein Barr virus (EBV) is linked to the pathogenesis of PTLD in the transplant population. Reports of EBV replication are as high as 80% in patients receiving antilymphocyte antibody drugs, with estimates of 20-30% for all patients on immunosuppressive therapy. PTLD, usually presenting as B cell lymphoproliferation, can range in severity from benign to rapidly fatal lymphoma. While antilymphocyte antibody drugs are implicated in reactivation of latent EBV, nucleotide synthesis inhibitors result in loss of recipient surveillance functions against B cells. Some additional risk of PTLD is conferred from preexisting EBV infection and CMV infection. While patients with a benign form of PTLD respond well to reduction of immunosuppression, PTLD B cell lymphoma is generally resistant to treatment.

MALIGNANCY

As is the case with incidence of infection in transplant recipients, the pathogenesis of malignancy reflects each patient's individual immunosuppression. The risk of immunosuppression-related malignancy is greatest in patients receiving several courses of rejection therapy with polyclonal anti-T-cell antibodies. Nonmelanotic skin and lip cancer are the most common types of tumors, with the incidence increasing with time after transplant. Patients are screened for the presence of these skin lesions and educated on the increased risk of UV exposure following transplantation.

Oncogenic viruses are more common in the immunocompromised host, and include human papilloma virus, associated with the development of carcinomas of the cervix, vulva and perineum, and hepatitis B and C, linked to hepatocellular carcinoma.

OTHER SYSTEMIC COMPLICATIONS OF IMMUNOSUPPRESSION

Many renal transplant recipients present for transplantation with essential hypertension. In addition, the use of immunosuppressive medications has been linked to secondary hypertension, often responding to a reduction in CsA and or steroid dose. Antihypertensive medications used in transplant recipients are discussed in Chapter 6 (page 125).

Hyperlipidemia is another recognized cardiovascular risk factor in renal transplant recipients, who tend to have an accelerated progression of the atherosclerotic process after transplantation. This is most likely due to immunosuppression, especially CsA, since tacrolimus and MMF do not appear to affect lipoprotein metabolism. While overall reduction in cardiovascular events with the use of lipid-lowering agents in renal transplantation has not been fully investigated, reduction in serum cholesterol promotes improved cardiovascular health. Antihyperlipidemic therapy is added to a program of diet surveillance and regular exercise in patients with multiple cardiovascular risk factors and LDL cholesterol exceeding 130 mg/dl, and in all kidney/pancreas recipients. The hydroxymethyl glutaryl coenzyme A (HMG CoA) reductase inhibitors (i.e., atorvastatin, Lipitor®) have been shown to be effective in reducing LDL cholesterol. Use of these agents should be conservative and monitored with frequent liver function studies. Cyclosporine prolongs the half life of statin drugs, leading to dosing in immunosuppressed patients at approximately 50% of manufacturer recommended dose (up to 80 mg daily atorvastatin). In addition, drugs of this class have been implicated in rhabdomyolosis and acute renal failure when given with CsA.

CONDITIONS UNIQUE TO STEROID ADMINISTRATION

Posttransplant glucose intolerance, also presenting as posttransplant diabetes mellitus (PTDM) is estimated to occur in up to 20% of patients. Steroids are primarily implicated in diabetes induction. Moderate hyperglycemia is treated with oral hypoglycemic agents (glyburide, glipizide), with a large portion of these patients eventually requiring insulin therapy. As steroid doses are reduced, insulin requirements lessen, and some patients regain the ability to autoregulate glucose metabolism.

Posttransplant bone loss can be very aggressive, especially in patients requiring frequent, high dose steroid therapy. Currently available antiosteoporosis agents are not recommended in many renal transplant patients due to the method of excretion and the impact on renal function. In addition, the major side effect of oral antiosteoporotic drugs is gastrointestinal, impacting absorption and metabolism of immunosuppressive medications. New synthetic calcitonin preparations, delivered as intranasal preparations, may hold promise in transplant recipients, but have not been widely used at this time.

OVER-THE-COUNTER MEDICATIONS

Due to the complexity of the posttransplant medication protocol, the use of over-the-counter medications, homeopathic preparations or megavitamins are discouraged until verified by the physician or transplant coordinator. Patients should be instructed to carry a list of their current medications and dosages with them at all times and to discuss all proposed self medication with their transplant physician or coordinator. The following common categories of over-the-counter medications are meant to demonstrate generally accepted choices, and should not be construed as blanket recommendations for all patients.

Acetaminophen (Tylenol®) is generally a safe analgesic choice for headaches or mild pain. Nonsteroidal anti-inflammatory agents (NSAIDS) (i.e., ibuprofen), should be avoided since they may precipitate acute renal failure in patients who are dependent on renal prostaglandins for maintenance of renal blood flow. Aspirin containing products and NSAIDS may lead to platelet disruption, possible increase in blood pressure or exacerbation of steroidal GI toxicity.

Famotidine (Pepcid®) may be utilized for stress ulcer prophylaxis postoperatively, with therapy often continued in patients with underlying ulcer disease or dyspepsia. Many transplant centers employ famotidine as part of the prescription drug protocol. Patients should avoid self medication with Cimetidine (Tagamet®), due to its indirect effect on CsA metabolism. Cimetidine has a known inhibitory effect on the metabolism of many drugs also known to interact with CsA.

Systemic decongestants cause an increase in blood pressure and heart rate, and are not routinely recommended in transplant patients. A short course (< 3 days) of topical decongestants (i.e., Afrin®, Neo-Synephrine®) is preferred, due to minimal absorption. Cough suppressants containing dextromethorphan (Robitussin DM®) can be used safely. Antihistamines (Benadryl®) are safe for short term treatment of allergy symptoms.

A summary of drugs commonly encountered in transplantation is available in the *Transplantation Drug Manual* (3rd edition, Pirsch, Simmons and Sollinger, Landes Bioscience, 1999).

CONCLUSION

As immunosuppressive medications continue to improve in efficacy and specificity, transplant recipients are living longer with well-functioning allografts. Complications of immunosuppression are many and can be severe. Strategies have been designed to emphasize prophylaxis and preemptive methods over therapeutic treatment. With the continued sophistocation of diagnostic methods and the current emphasis on establishing cellular tolerance of allografts, reliance on immunosuppression may decrease in the coming years. Until that time, however, successful outcome of transplantation continues to be a delicate balance between immunosuppression and maintenance of a baseline immune response toward invading pathogens. Optimizing early graft function lessens metabolic and systemic complications of overimmunosuppression, increasing the chances of graft function and patient longevity.

SELECTED READING

1   Amonte AJ, Kahan BD. Abbreviated area-under-the-curve strategy for monitoring cyclosporine microemulsion therapy in immediate posttransplant period. Clin Chem 1996; 42(8 pt 1):1294-6.

12

2   Arnadottir M, Berg AL. Treatment of hyperlipidemia in renal transplant recipients. Transpl 1997; 63:339-45.

3   Bumgardner GL, Matas AJ. Transplantation and pregnancy. Transpl Rev 1992; 6:139-62.

4   Danovitch GM. Immunosuppressive Medications and Protocols for Kidney Transplantation. In: Danovitch GM, ed. Handbook of Kidney Transplantation. Boston: Little, Brown & Co, 1996:55-93.

5   First MR. An update on the new immunosuppressive drugs undergoing preclinical and clinical trials: potential applications in organ transplantation. Am J Kid Dis 1997; 29:303-17.

6   Hibberd PL, Rubin RH. Clinical aspects of fungal infections in organ transplant recipients. Clin Inf Dis 1994; 19:33S-40S.

7   Hooks MA. Tacrolimus, a new immunosuppressant–a review of the literature. Ann Pharmacother 1994; 28:501-11.

8   Hricik DE, Almawi WY, Strom TB. Trends in the use of glucocorticoids in renal transplantation. Transpl 1994; 57:979-89.

9   Jindal RM. Posttransplant diabetes mellitus–a review. Transpl 1994; 58:1289-98.

10   Kahan BD. Reduction of the occurrence of acute cellular rejection among renal allograft recipients treated with Basiliximab, a chimeric anti-interleukin-2-receptor monoclonal antibody. United States Simulect Renal Study Group. Transpl 1999; 67(2), 276-84.

11   Kahan BD. Sirolimus: a new agent for clinical renal transplantation. Transpl Proc 1997; 29:48-50.

12   Kamada H, Itoh H, Shibata H, et al. Inhibitory mechanism of mizorbine on the antibody production of mouse B cells stimulated with lipopolysaccharide. Jpn J Pharmacol 1997; 74(4), 323-30.

13   Knoop M, Neumann U, Langrehr JM, Neuhaus P. Brequinar Sodium-Properties of a new immunosuppressive drug. TxMed 1994; 3:223.

14   Lake KD. Management of drug interactions with cyclosporine. Pharmacother 1991; 11(5):110S-18S.

15   Lebreton L, Annat J, Derrepas P, Dutartre P, Renaut P. Structure-immunosuppressive activity relationships of new analogues of 15-deoxyspergualin. 1. Structural modifications of the hydroglycine moiety. J Med Chem 1999; 42(2): 277-90.

16   Noble S, Markham A. Cyclosporine: a review of the pharmacokinetic properties, clinical efficacy and tolerability of a microemulsion-based formulation (Neoral). Drugs 1995; 50:924-41.

17   Penn I. The changing pattern of posttransplant malignancies. Transpl Proc 1991; 23:1101-3.

18   Rubin RH, Tolkoff-Rubin NE. The impact of infection on the outcome of transplantation. Transpl Proc 1991; 23:2068-74.

19   Silva Junior HT, Morris RE. Leflunomide and malononitrilamides. Am J Med Sci 1997; 313(5): 289-301.

20   Stiller CR. An overview of the first decade of cyclosporine. Transpl Proc 1996; 8:2005-12.

21   Suthanthiran M, Morris RE, Strom TB. Immunosuppressants: cellular and molecular mechanisms of action. Am J Kid Dis 1996; 28:159-72.

22   The Tricontinental Mycophenolate Mofetil Renal Transplantation Study Group. A blinded, randomized clinical trial of mycophenolate mofetil for the preven-

tion of acute rejection in cadaveric renal transplantation. Transpl 1996; 61:1029-31.

23   Xu X, Williams JW, Shen J, et al. In vitro and in vivo mechanisms of action of the antiproliferative and immunosuppressive agent, brequinar sodium. J Immunol 1998; 160(2): 846-53.

24   Fishman JA, Rubin RH. Medical progress: Infection in organ transplant recipients. New England Journal of Medicine 1998; 338(24): 1741-51.

25   Pirsch J, Simmons W, Sollinger H. Transplantation Drug Manual, 3rd edition. Austin: Landes Bioscience, 1999.

12

Index

Index

Index